RealTime Physics
Active Learning Laboratories

Module 3

Electricity and Magnetism

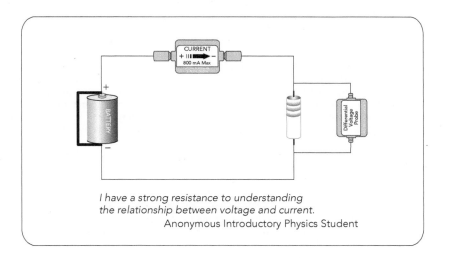

I have a strong resistance to understanding the relationship between voltage and current.
Anonymous Introductory Physics Student

David R. Sokoloff
Department of Physics
University of Oregon

Priscilla W. Laws
Department of Physics
Dickinson College

with contributions by

Ronald K. Thornton
Departments of Physics and Education
Tufts University

John Wiley & Sons, Inc.

EXECUTIVE EDITOR	Stuart Johnson
ASSOCIATE EDITOR	Alyson Rentrop
MARKETING MANAGER	Christine Kushner
SENIOR PRODUCTION EDITOR	Sujin Hong
PRODUCTION MANAGEMENT SERVICES	Aptara, Inc.
PHOTO RESEARCHER	Sheena Goldstein
COVER DESIGNER	Kristine Carney
COVER PHOTO	© CTR design LLC/iStockphoto

This book was set in Palatino by Aptara/Delhi and printed and bound by Quad/Graphics. The cover was printed by Quad/Graphics.

This book is printed on acid free paper. ∞

Founded in 1807, John Wiley & Sons, Inc. has been a valued source of knowledge and understanding for more than 200 years, helping people around the world meet their needs and fulfill their aspirations. Our company is built on a foundation of principles that include responsibility to the communities we serve and where we live and work. In 2008, we launched a Corporate Citizenship Initiative, a global effort to address the environmental, social, economic, and ethical challenges we face in our business. Among the issues we are addressing are carbon impact, paper specifications and procurement, ethical conduct within our business and among our vendors, and community and charitable support. For more information, please visit our website: www.wiley.com/go/citizenship.

Evaluation copies are provided to qualified academics and professionals for review purposes only, for use in their courses during the next academic year. These copies are licensed and may not be sold or transferred to a third party. Upon completion of the review period, please return the evaluation copy to Wiley. Return instructions and a free of charge return mailing label are available at www.wiley.com/go/returnlabel. If you have chosen to adopt this textbook for use in your course, please accept this book as your complimentary desk copy. Outside of the United States, please contact your local sales representative.

ISBN 978-0-470-76889-1

Printed in the United States of America

V10018750_052120

PREFACE

Development of the series of *RealTime Physics (RTP)* laboratory guides began in 1992 as part of an ongoing effort to create high-quality curricular materials, computer tools, and apparatus for introductory physics teaching.[1] The *RTP* series is part of a suite of *Activity-Based Physics* curricular materials that include the *Tools for Scientific Thinking* laboratory modules,[2] the *Workshop Physics* Activity Guide,[3, 4] and the *Interactive Lecture Demonstration* series.[5, 6] The development of all of these curricular materials has been guided by the outcomes of physics education research. This research has led us to believe that students can learn vital physics concepts and investigative skills more effectively through guided activities that are enhanced by the use of powerful microcomputer-based laboratory (MBL) tools.

MBL tools—originally developed at Technical Education Research Centers (TERC) and at the Center for Science and Mathematics Teaching, Tufts University— have become increasingly popular for the real-time collection, display, and analysis of data in the introductory laboratory. MBL tools consist of electronic sensors, a microcomputer interface, and software for data collection and analysis. Sensors are available to measure such quantities as force, sound, magnetic field, current, voltage, temperature, pressure, rotary motion, acceleration, humidity, light intensity, pH, and dissolved oxygen.

MBL tools provide a powerful way for students to learn physics concepts. For example, students who walk in front of an ultrasonic motion sensor while the software displays position, velocity, and/or acceleration in real time more easily discover and understand motion concepts. They can see a cooling curve displayed instantly when a temperature sensor is plunged into ice water, or they can sing into a microphone and see a pressure vs. time plot of sound intensity.

MBL data can also be analyzed quantitatively. Students can obtain basic statistics for all or a selected subset of the collected data, and then either fit or model the data with an analytic function. They can also integrate, differentiate, or display a fast Fourier transform of data. Software features enable students to generate and display *calculated quantities* from collected data in real time. For example, since mechanical energy depends on mass, position, and velocity, the time variation of potential and kinetic energy of an object can be displayed graphically in real time. The user just needs to enter the mass of the object and the appropriate energy equations ahead of time.

The use of MBL tools for both conceptual and quantitative activities, when coupled with developments in physics education research, has led us to expand our view of how the introductory physics laboratory can be redesigned to help students learn physics more effectively.

COMMON ELEMENTS IN THE *REALTIME PHYSICS* SERIES

Each laboratory guide includes activities for use in a series of related laboratory sessions that span an entire quarter or semester. Lab activities and homework assignments are integrated so that they depend on learning that has occurred during the previous lab session and also prepare students for activities in the next session. The major goals of *RealTime Physics* are (1) to help students acquire an understanding of a set of related physics concepts; (2) to provide students with direct experience of the physical world by using MBL tools for real-time data collection, display, and analysis, (3) to enhance traditional laboratory skills; and (4) to reinforce topics covered in lectures and readings using a combination of conceptual activities and quantitative experiments.

To achieve these goals we have used the following design principles for each module based on educational research:

- The materials for the weekly laboratory sessions are sequenced to provide students with a coherent observational basis for understanding a single topic area in one semester or quarter of laboratory sessions.

- The laboratory activities invite students to construct their own models of physical phenomena based on observations and experiments.

- The activities are designed to help students modify common preconceptions about physical phenomena that make it difficult for them to understand essential physics principles.

- The activities are designed to work best when performed in collaborative groups of 2 to 4 students.

- MBL tools are used by students to collect and graph data in real time so they can test their predictions immediately.

- A learning cycle is incorporated into each set of related activities that consists of prediction, observation, comparison, analysis, and quantitative experimentation.

- Opportunities are provided for class discussion of student ideas and findings, if lab scheduling affords that possibility.

- Each laboratory comes with a pre-lab warm-up assignment, and with a post-lab homework assignment that reinforces critical physics concepts and investigative skills.

The core activities for each laboratory session are designed to be completed in two hours. Extensions have been developed to provide more in-depth coverage when longer lab periods are available. The materials in each laboratory guide are comprehensive enough that students can use them effectively even in settings where instructors and teaching assistants have minimal experience with the curricular materials.

The curriculum has been designed for distribution in electronic format. This allows instructors to make local modifications and reprint those portions of the materials that are suitable for their programs. The *Activity-Based Physics* curricular materials can be combined in various ways to meet the needs of students and instructors in different learning environments. The *RealTime Physics* laboratory guides are designed as the basis for a complete introductory physics laboratory program at colleges and universities, but they can also be used as the central component of a high school physics course. In a setting where formal lectures are given we recommend that the *RTP* laboratories be used in conjunction with *Interactive Lecture Demonstrations*.[5, 6]

THE ELECTRICITY AND MAGNETISM LABORATORY GUIDE

The primary goal of this *RealTime Physics Electricity and Magnetism* guide is to provide students with a solid understanding of basic electricity and magnetism concepts (including DC circuits). A number of physics education researchers have documented that most students begin their studies with conceptions about the nature of these concepts that can seriously inhibit their learning.[7, 8]

For example, McDermott and Shaffer[8] have documented the following student difficulties with circuits, among others: (1) failure to distinguish among concepts of current, potential difference, energy, and power; (2) lack of concrete experiences with real circuits; (3) failure to understand and apply the concept of a complete circuit; (4) belief that direction of current and order of elements matter; (5) belief that current is "used up" in a circuit; (6) belief that the battery is a constant current source; (7) failure to recognize that an ideal battery maintains a constant potential between its terminals; (8) failure to distinguish between branches connected in parallel across a battery and branches connected in parallel elsewhere; (9) failure to distinguish between potential and potential difference; and (10) difficulty in identifying series and parallel connections. This research was used in the creation of activities for *Physics by Inquiry*[9], which were modified and expanded for *Workshop Physics, Module 4*[10], on which the circuit labs in this manual are based. The Electricity and Magnetism materials in this manual (Labs 1, 2, 3, 9 &10) were based in part on *Module 4* of the *Workshop Physics Activity Guide*[10], on *Explorations in Physics* Unit H[11], as well as on the Faraday's Law Activity in *Physics with Video Analysis*.[12]

RealTime Physics Electricity and Magnetism includes 10 labs:

Lab 1 (Electric Charges, Forces and Fields):   Students first examine the hypothesis that there are two types of charge, by looking at the interactions of strips of Scotch Magic© tape charged in various ways. Then interactions of aluminum foil covered and uncovered Styrofoam cups, charged by induction, are used to test the hypothesis that certain materials are conductors of electric charge, while others are insulators. Coulomb's law is examined quantitatively by analyzing a video of a charged prod brought near a hanging charged ball. Finally, the electric field around the charged tip of a rod is examined using a charged strip of Scotch Magic© tape as a test charge.

Lab 2 (Electric Fields, Flux and Gauss' Law):   The representation of the electric field using field lines is examined using a computer simulation of the field of a number of simple charge distributions. Electric flux is defined. Its dependence on the angle between the normal to the surface and the direction of the field is examined quantitatively using a model fabricated from nails on a grid. Then Gauss' law in "flatland" is examined by drawing "two-dimensional" surfaces around different charge distributions. Finally, the implication of Gauss' law regarding the location of excess charge on a conductor is examined with a version of Faraday's ice pail experiment.

Lab 3 (Electric and Gravitational Potential):   The gravitational and electrostatic forces are compared in mathematical form and magnitude. Then the meaning of work being independent of path is examined for the conservative gravitational force using a force sensor and low-friction cart. Potential difference is defined. Then, the equipotentials for several simple charge distributions are found using carbonized paper with electrodes painted using conducting paint.

Lab 4 (Batteries, Bulbs, and Current):   In this lab students first explore the nature and definition of electric current. They then discover what is necessary for current to flow in a complete circuit, and practice connecting a variety of circuits. They use current sensors to determine a model for the current flowing in a simple series circuit, and current and voltage sensors to examine the potential difference maintained by a battery for various currents drawn from the battery.

Lab 5 (Current in Simple DC Circuits):   Students explore the relationships between current in different parts of series and parallel circuits constructed first from bulbs and then from resistors. They use current and voltage sensors to do their measurements.

Lab 6 (Voltage in Simple DC Circuits and Ohm's Law): Students explore the relationships between potential difference in different parts of series and parallel circuits constructed from bulbs and resistors. They also examine the internal resistance of a battery, and discover Ohm's law for a resistor by graphing the current through a resistor and the voltage across it simultaneously. They use current and voltage sensors to do their measurements.

Lab 7 (Kirchhoff's Circuit Rules): First students examine how a multimeter is connected to measure current and voltage, and why it is connected that way. Then they learn how to measure resistance with a multimeter, and they use the multimeter to discover the rules for finding the equivalent resistance for series and parallel connections of resistors. Finally, they apply Kirchhoff's Circuit Rules to more complex circuits.

Lab 8 (Introduction to Capacitors and RC Circuits): Students first construct parallel-plate capacitors from aluminum foil sheets and examine the dependence of the capacitance on plate separation and plate area. Then they discover the rules for finding the equivalent capacitance for series and parallel connections of capacitors. Finally, they explore the transient behavior of RC circuits and the definition of time constant for an exponential decay.

Lab 9 (Magnetism): Magnetism is explored with permanent magnets and various objects. The orientation behavior of a compass needle is examined. Ceramic disk magnets are used to examine the existence of N and S poles only in pairs. The magnetic field of a rod-shaped magnet is examined using compasses, iron filings and a magnetic field sensor. Magnetization is explored using paper clips and a permanent bar magnet. The magnetic forces on charges at rest and moving are examined using Scotch Magic© tape and a cathode ray tube. Finally, the Lorentz Force law is compared to the observations of magnetic forces, and the magnetic force on a current-carrying wire is observed.

Lab 10 (Electromagnetism): The magnetic field of a current-carrying wire is examined using compasses and a magnetic field sensor. Then the interactions between a rod-shaped magnet and a coil of wire are examined qualitatively. Magnetic flux is defined, and qualitative observations are used to arrive at Faraday's law. Then, emfs induced in a coil by the movement of a rod-shaped magnet are examined quantitatively by analyzing videos of the motion of the magnet relative to the coil. Finally, Lenz's law is explored by analyzing the direction of the current induced by a magnet moving relative to a coil.

ON-LINE TEACHERS' GUIDE

The *Teachers' Guide* for *RealTime Physics Electricity and Magnetism* is available on-line at **http:/www.wiley.com/college/sokoloff**. This *Guide* focuses on pedagogical (teaching and learning) aspects of using the curriculum, as well as computer-based and other equipment. The *Guide* is offered as an aid to busy physics educators and does not pretend to delineate only the "right" way to use the *RealTime Physics Electricity and Magnetism* curriculum and certainly not the MBL tools. There are many right ways. The *Guide* does, however, explain the educational philosophy that influenced the design of the curriculum and tools and suggests effective teaching methods. Most of the suggestions have come from the college, university, and high school teachers who have participated in field testing of the curriculum.

The *On-line Teachers' Guide* has eleven sections. Section I presents suggestions regarding computer hardware and software to aid in the implementation of this activity-based MBL curriculum. Sections II through XI present information about the ten laboratories. Included in each of these is information about the specific equipment and materials needed, tips on how to optimize student learning, answers to questions in the labs, and complete answers to the homework.

EXPERIMENT CONFIGURATION FILES

Experiment configuration files are used to set up the appropriate software features to go with the activities in these labs. You will need the set of files that is designed for the software package you are using, or you will need to set up the files yourself. At this writing, experiment configuration files for *RealTime Physics Electricity and Magnetism* are available for the Vernier Software and Technology *Logger Pro* (for Windows and Macintosh) and PASCO *Data Studio* (for Windows and Macintosh). Appendix A of this module outlines the features of the experiment configuration files for *RealTime Physics Electricity and Magnetism,* as a guide to setting up configuration files for other software packages. For more information, consult the *On-line Teachers' Guide.*

CONCLUSIONS

RealTime Physics Electricity and Magnetism has been used in a variety of different educational settings. Many university, college, and high school faculty who have used this curriculum have reported improvements in student understanding of *electricity and magnetism* concepts. These comments are supported by our careful analysis of pre- and post-test data for example using the *Electric Circuit Conceptual Evaluation,* some of which have been reported in the literature.[13] Similar research on the effectiveness of *RealTime Physics Mechanics,*[14] *Heat and Thermodynamics,* and *Light and Optics* also shows dramatic conceptual learning gains in these topic areas. We feel that by combining the outcomes of physics educational research with microcomputer-based tools, the laboratory can be a place where students acquire a mastery of both difficult physics concepts and vital laboratory skills.

ACKNOWLEDGMENTS

RealTime Physics Module 3: Electricity and Magnetism could not have been developed without the hardware and software development work of Ronald Thornton, David Vernier, Stephen Beardslee, Lars Travers, and Ronald Budworth. We thank David Jackson (Dickinson College) and Kerry Browne (Rivendell Academy) for their permission to adapt magnetism activities from *Explorations in Physics*[15] for use in Lab 9 and Robert Teese (Rochester Institute of Technology) for his contributions to the video analysis activities in Labs 1 and 10. Patrick Cooney (Millersville University) also contributed to the design of the Lab 10 video activities. We are indebted to numerous college, university, and high school physics teachers, and especially Curtis Hieggelke (Joliet Junior College), John Garrett (Sheldon High School, retired), and Maxine Willis (Gettysburg High School, retired) for beta testing earlier versions of the laboratories with their students. We thank Mary Fehrs, Matthew Moelter, Gene Mosca, Sharon Schmalz, Tom Carter and Greg Putnam for beta testing the circuits labs and for their invaluable suggestions and corrections in the final stages of editing.

At the University of Oregon, we especially thank Dean Livelybrooks for supervising the introductory physics laboratory, for providing invaluable feedback, and for writing some of the homework solutions for the *Teachers' Guide*. Frank Womack, Dan DePonte, Sasha Tavenner, and all of the introductory physics laboratory teaching assistants provided valuable assistance and input. We also thank the faculty at the University of Oregon (especially Stan Micklavzina), Tufts University, Lane Community College (especially Dennis Gilbert) and Dickinson College for their input, and for assisting with our conceptual learning assessments. Finally, we could not have even started this project if not for our students' active participation in these endeavors.

This work was supported in part by the National Science Foundation under grant number DUE-9455561, *"Activity Based Physics: Curricula, Computer Tools, and Apparatus for Introductory Physics Courses,"* grant number USE-9150589, *"Student Oriented Science,"* grant number DUE-9451287, *"RealTime Physics II: Active University Laboratories Based on Workshop Physics and Tools for Scientific Thinking,"* grant number USE-9153725, *"The Workshop Physics Laboratory Featuring Tools for Scientific Thinking,"* and grant number TPE-8751481, *"Tools for Scientific Thinking: MBL for Teaching Science Teachers,"* and by the Fund for Improvement of Post-secondary Education (FIPSE) of the U.S. Department of Education under grant number G008642149, *"Tools for Scientific Thinking,"* and number P116B90692, *"Interactive Physics."*

REFERENCES

1. David R. Sokoloff, Ronald K. Thornton and Priscilla W. Laws, "RealTime Physics: Active Learning Labs Transforming the Introductory Laboratory," *Eur. Journal of Phys.* **28:** S83-S94 (2007).

2. Ronald K. Thornton and David R. Sokoloff, "Tools for Scientific Thinking— Heat and Temperature Curriculum and Teachers' Guide," (Portland, Vernier Software, 1993) and David R. Sokoloff and Ronald K. Thornton, "Tools for Scientific Thinking—Motion and Force Curriculum and Teachers' Guide," 2nd ed. (Portland, Vernier Software, 1992).

3. P. W. Laws, "Calculus-based Physics Without Lectures," *Phys. Today* **44** (12): 24–31 (1991).

4. Priscilla W. Laws, *Workshop Physics Activity Guide: The Core Volume with Module 1: Mechanics, Second Edition*, (Hoboken, NJ, Wiley, 2004).

5. David R. Sokoloff and Ronald K. Thornton, "Using Interactive Lecture Demonstrations to Create an Active Learning Environment," *The Physics Teacher* **27** (6): 340 (1997).

6. David R. Sokoloff and Ronald K. Thornton, *Interactive Lecture Demonstrations*, (Hoboken, NJ, Wiley, 2004).

7. L. C. McDermott, "Millikan Lecture 1990: What We Teach and What Is Learned—Closing the Gap," *Am. J. Phys* **59:** 301 (1991).

8. Lillian C. McDermott and Peter S. Shaffer, "Research as a Guide to Curricular Development: An Example from Introductory Electricity. Part I: Investigation of Student Understanding," *Am. J. Phys.* **60:** 994 (1992).

9. Lillian C. McDermott, *Physics by Inquiry*, (Hoboken, NJ, Wiley, 1996).

10. Priscilla W. Laws, *Workshop Physics Activity Guide: Module 4, Electricity and Magnetism, Second Edition* (Hoboken, NJ, Wiley, 2004).

11. D. Jackson, & K. Browne, *Explorations in Physics,* Unit H—Magnets, Charge and Electric Motors can be downloaded at http://physics.dickinson.edu/~eip_web/Resources_UnitH_ActivityGuide.html.

12. P.W. Laws, et. al., *Physics with Video Analysis* (Activity 32), (Portland OR, Vernier, 2009).

13. David R. Sokoloff, "Teaching Electric Circuit Concepts Using Microcomputer-based Current and Voltage Sensors," chapter in *Microcomputer-Based Labs: Educational Research and Standards*, Robert F. Tinker, ed., *Series F, Computer and Systems Sciences*, **156,** 129–146 (Berlin, Springer-Verlag, 1996).

14. Ronald K. Thornton and David R. Sokoloff, "Assessing Student Learning of Newton's Laws: The *Force and Motion Conceptual Evaluation* and the Evaluation of Active Learning Laboratory and Lecture Curricula," *Am. J. Phys.* **66:** 338–352 (1998).

15. David P. Jackson, Priscilla W. Laws and Scott V. Franklin, *Explorations in Physics*, (Hoboken, NJ, Wiley, 2003).

This project was supported, in part, by the National Science Foundation. Opinions expressed are those of the authors and not necessarily those of the foundation.

Contents

Contents

Name_____   Date_____

Pre-Lab Preparation Sheet for
Lab 1—Electric Charges, Forces, and Fields

(Due at the beginning of lab)

Directions: Read over Lab 1 and then answer the following questions about the procedures.

1. Describe briefly what types of observations you will make in Activity 1-1 to determine whether like or unlike charges attract each other.

2. What do you predict will happen to the force between two charges when the distance between them decreases? How will you test this qualitatively?

3. What do you predict will happen to the force between two charges when the sign of one of the charges is changed?

4. What do you predict will happen when a *charged* foam cup is brought near an *uncharged,* aluminum foil-covered cup?

5. Describe briefly how you will explore the direction of the electric field produced by a positively charged rod in Activity 3-1.

LAB 1:
ELECTRIC CHARGES, FORCES, AND FIELDS

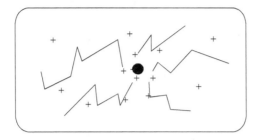

If anyone should doubt whether the electrical matter passes through the substance of bodies, or only over along their surfaces, a shock from an electrified large glass jar, taken through his own body, will probably convince him.

—Benjamin Franklin (1706–1790)

OBJECTIVES

- To discover some of the interactions of particles that carry electric charges.
- To understand how Coulomb's law describes the forces between charges.
- To understand the concept of electric fields.

OVERVIEW

On cold clear days rubbing almost any object seems to create a new kind of force. After being used, a plastic comb will pick up bits of paper, hair, or cork with it. (And when you think about it, this force is relatively strong. It is able to lift your hair up against the gravitational pull of the whole Earth downward on your hair!) Anyone who has walked across a carpet and then been shocked by touching a light switch or electrical appliance will attest to the presence of "electric forces." You are going to begin a study of electrical phenomena by exploring the nature of the forces between objects that have been rubbed or pulled apart, or come into contact with other objects that have had such interactions. These forces are attributed to a fundamental property of the constituents of atoms known as *charge*. The forces between charged particles that are not moving or are moving relatively slowly are known as *electrostatic forces*.

You can start your study by exploring the circumstances under which electrostatic forces are attractive or repulsive. This should allow you to establish how many types of charge there are. Then you can proceed to a qualitative study of how the force between charged objects depends on the distance between the charged objects. This will lead you to a formulation of *Coulomb's law*, the mathematical relationship that describes the vector force between two small charged objects. Next you can carry out a more quantitative experiment on the repulsion between two charged objects when they are brought closer and closer together.

Finally we will define a new quantity called *electric field* that can be used to determine the net force on a small test charge due to the presence of other charges at different locations. You will learn how to use Coulomb's law to calculate the electric field at various points of interest arising from a small number of charges.

INVESTIGATION 1: ELECTROSTATIC FORCES

Exploring the Nature of Electrical Interactions

You can investigate the properties of electrical interactions between objects using the following materials:

- roll of Scotch Magic© tape

- hard rubber, hard plastic or Teflon© rod and fur

- glass or acrylic rod and polyester, felt or silk cloth

Although the nature of electrical interactions is not obvious without careful experimentation and reasoning, let's start by considering a plausible hypothesis:

Hypothesis I: If the interaction between objects that have been rubbed or pulled apart is due to a property of matter called charge, then there are only two types of electrical charge. For the sake of convenience we will call these charges positive charge and negative charge.

Try the activities suggested below. Mess around and see if you can design careful, logical procedures to demonstrate that there must be at least two types of charge. Carefully explain your observations and state reasons for any conclusions you draw.

Note: In doing the following activities and answering the questions, you are not allowed to state previously memorized results. You must devise and describe a sound and logical set of observations that support or disprove Hypothesis I. In answering the questions, carefully describe the observations you used to reach your conclusions.

Activity 1-1: Test of Hypothesis I

1. You and your partner should each tape a 10 cm or so strip of Scotch Magic© tape onto the lab table. The end of each tape should be curled under to make a non-stick handle. Peel your tape off the table and bring the non-sticky side of the tape toward your partner's non-sticky side. Hold both strips vertically.

Question 1-1: Describe your observations. Do the strips attract, repel or not interact?

2. Tape two more strips of tape with "handles" on the table and use a pen to label them "B" for bottom. Press a second strip of tape on top of each of the B pieces, and also give it a handle. Label these strips "T" for top.

3. Pull each pair of strips off the table. Then pull each top and bottom strip apart. (**Note:** you will need to repeat this set of procedures several times to answer all the questions below.)

Question 1-2: Describe the interaction between two top (T) strips when they are brought near each other. Do the strips attract, repel or not interact at all?

Question 1-3: Describe the interaction between two bottom (B) strips when they are brought near each other. Do the strips attract, repel or not interact at all?

Question 1-4: Describe the interaction between a top (T) and a bottom (B) strip when they are brought near each other. Do the strips attract, repel or not interact at all?

Question 1-5: Are your observations of the tape strip interactions consistent with Hypothesis I, i.e., that there are two types of charge? Explain your answer carefully, in complete sentences, and using the results of *all* your observations.

Question 1-6: Do like charges repel or attract each other? Do unlike charges repel or attract each other? Explain based on your observations.

Question 1-7: Based on your observations of the movement of the tapes, how does the strength of these forces compare to the gravitational force on the tapes near the surface of the earth?

You know from your everyday experiences that when objects rub on each other, static charges can build up, e.g., the soles of your shoes rubbing on the carpet. In earlier times scientists transferred charge to objects by rubbing a rubber rod with fur or by rubbing a glass rod with silk. These days we usually use polyester instead of silk and hard plastic or Teflon instead of rubber. In the next activity you can continue studying the interactions between charged objects using techniques developed by early investigators.

Activity 1-2: Charged Rods

1. Tape a single strip of Scotch Magic© tape to the table as in Activity 1-1.

2. Rub a rubber, black plastic, or Teflon rod vigorously with fur.

3. Pull the tape off the table, hold it vertically, and bring the tip of the rod close to the tape without touching it.

Question 1-8: What happens to the tape when you bring the rod that has been rubbed with fur close to it?

4. Now repeat steps (1)–(3) using a glass or acrylic rod rubbed with polyester (or silk).

Question 1-9: Compare the interaction between the glass (or acrylic) rod and the tape to that between the black plastic or Teflon rod and tape in (3).

Question 1-10: Recalling the interactions between like and unlike charged objects that you observed with the Scotch Magic© tape in the previous activity, do these new observations add support to Hypothesis 1? Explain

> **Comment:** Benjamin Franklin *arbitrarily* assigned the term "negative" to the nature of the charge that results when a hard plastic, Teflon, or rubber rod is rubbed with fur. Conversely, the nature of the charge found on the glass or acrylic rod after it is rubbed with silk (or polyester) is defined as "positive." (The term "negative" could just as well have been assigned to the charge on the glass rod, or these charges could have been called "blue charge" and "red charge." The original choice of names was arbitrary.)

Question 1-11: Given the designation of the charge on a glass or acrylic rod as positive and the charge on a hard plastic, Teflon, or rubber rod as negative, and your observations with pieces of tape, what is the sign of the charge on the Scotch Magic© tape when it is pulled up from the tabletop? Explain how you know.

Question 1-12: What are the signs of the charges on the "B" and "T" pieces of Scotch Magic© tape in Activity 1-1? Explain how you know.

Hypothesis II: Conductors and Non-conductors

There is a second hypothesis that we will consider that has to do with the properties of materials. Scientists believe that most matter is made up of atoms that contain positive and negative charges associated with protons and electrons, respectively. When electrons in an atom surround an equal number of protons the charges neutralize each other and the atom does not interact with other charges outside the solid. In some types of solid materials, known as *insulators*, the electrons are tightly bound to the protons in the atoms and do not move away from their atoms. However, in other solids known as *conductors*, the electrons—but not the protons—are free to move under the influence of other charges.

Hypothesis II: Charge moves readily on certain materials, known as conductors, and not on others, known as insulators. In general, metals are good conductors, while glass, rubber, and plastic tend to be insulators.

To test Hypothesis II you will need the materials used before along with

- two Styrofoam cups, one completely covered on the outside with aluminum foil.

- non-conducting string

- a stand with a vertical rod about 0.8–1.0 m long, with a clamp and cross rod

- a third Styrofoam cup attached at the end of a plastic straw

- a nitrile disposable glove

Activity 1-3: Insulators and Conductors

1. Hang the two upside-down Styrofoam cups from the rod, at the end of strings so that they each hang about an inch above the table. The cups should be several inches apart.

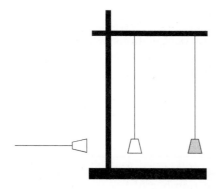

2. Put on the glove, and charge the cup on the straw by rubbing the cup vigorously with your gloved hand. Bring the cup near but not touching the aluminum foil-covered Styrofoam cup. What happens?

3. Recharge the cup on the straw, and repeat observation (2) by bringing it near the uncovered Styrofoam cup.

Question 1-13: Compare your observations with the aluminum foil-covered and uncovered cups. Describe what happened. Was either affected by the charged cup on the straw? If both were affected, which one was affected more?

Question 1-14: Use your knowledge of how like and unlike charges interact with each other, and your observations to explain how this activity supports Hypothesis II. What do you think happened to the charges on the aluminum foil-covered cup when the charged cup was brought near it?

Note: Touching a charged object with your finger allows some charge to flow off (or onto) the object, through your body to the ground (or from the ground to the object). This can be an effective way of discharging (or depositing a charge on) the object.

---

4. Charge the cup on the straw as before, and again bring it near, but not touching the uncovered Styrofoam cup.

5. With the rod held steady, have a partner touch a finger gently to the side of the cup to discharge it.

6. Repeat (4) and (5) with the aluminum foil-covered Styrofoam cup. Again bring the charged cup on the straw near, but not touching the foil-covered Styrofoam cup. Then gently touch the side of the cup with a finger.

Question 1-15: Describe what happened with each cup. Was there any difference in the behaviors?

Question 1-16: Use your observations when the aluminum-covered and non-covered cups were touched with a finger, and the note above about charge flowing to or from ground through a finger, to support Hypothesis II.

Comment: We picture "uncharged" objects made up of a huge number of atoms having an equal number of negatively charged electrons that swarm around positively charged protons that are inside the atomic nuclei. These two types of charged particles neutralize each other. On the other hand, a "charged object" has an excess of either electrons or protons. For this reason, we refer to a charged object as having "an excess charge" or "a net charge."

The main difference between conductors and insulators is that conductors (like the aluminum foil) have some electrons that are relatively free to move around. In insulators the electrons are bound to the protons, and cannot move

---

very much. If the cup on the straw is charged negatively (cup with excess negative charge on it) and is brought near the aluminum foil-covered cup, it can rearrange the "free" electrons by repelling them to the opposite side of the aluminum foil-covered cup. Then it can attract the positive charge remaining on the side nearest it. This separation of charge is called *polarization*. If your finger touches the aluminum foil, it allows electrons to flow off, leaving the foil with a net positive charge. This charging of the aluminum foil-covered cup without touching the charged cup on the straw to it is called charging by *induction*.

When the charged cup on the straw is brought toward the uncovered Styrofoam cup (insulator), it can only displace the electrons a little bit, so the attraction is much less. This process is still called *polarization*, but the effect is much smaller.

INVESTIGATION 2: FORCES BETWEEN CHARGED PARTICLES—COULOMB'S LAW

Coulomb's law is a mathematical description of the fundamental nature of the electrical forces between charged objects that are small compared to the distance between them (so that they act more or less like point particles). Coulomb's law is usually stated without experimental proof in most introductory physics textbooks. Instead of just accepting the textbook statement of Coulomb's law, you are going to first examine qualitatively how the force between two charged objects depends on their separation. Then, you will make measurements using a video of two small charged spheres, enabling you to determine the forces between them quantitatively.

First a prediction.

Prediction 2-1: How do you think the magnitude of the force between two charged objects will change as you change the distance between the objects? What will happen to the force if you decrease the distance? What will happen to the force if you increase the distance?

To test your prediction, you will need

- roll of Scotch Magic© tape
- glass or acrylic rod and polyester or silk cloth

Activity 2-1: Qualitative Look at Coulomb's Law

1. Attach a 10 cm or so length of Scotch Magic© tape to the tabletop with a small "handle" on one end as you did in Activity 1-1.

2. Charge the glass or acrylic rod by rubbing it with the cloth.

3. Pull the tape off the table, and observe the force exerted on the tape when the end of the rod is brought closer to and further from the tape. Do not let the tape touch the rod.

Question 2-1: What seems to happen to the force of interaction between the charged tape and charged rod as the distance between them *decreases*?

Question 2-2: On the basis of the observations you have already made, does the force between the two charged objects seem to lie along a line between them or in some other direction? Explain. **Hint:** What would happen to the mutual repulsion or attraction if the force did not lie on a line between the objects?

Quantitative Look at Coulomb's Law

In the late eighteenth century, Charles-Augustin de Coulomb used an elaborate torsion balance, and a great deal of patience, to verify that the force of interaction between small spherical charged objects varied as the inverse square of the distance between them. Verification of the inverse square law can also be accomplished using modern apparatus.

A small, conducting sphere can be placed on the end of an insulating rod and then negatively charged using a plastic or Teflon rod that has been rubbed with fur. This charged sphere can be used as a prod to cause another charged sphere, suspended from a thread (actually two threads, to keep it stable), to rise to a larger and larger angle as the prod comes closer, as shown in the diagram below.

As you have seen in previous activities, carrying out electrostatics experiments is difficult, and it is hard to get good quantitative results. Reasonable measurements can be made by using a video camera under fairly ideal conditions to record how the hanging ball moves as the prod comes closer and closer to it. You can then take measurements directly from the video frames. With modern software, this is relatively easy to do. You can carry out the experiment by analyzing a video that has already been made for you.

Activity 2-2: Forces on a Suspended Charged Object—Theory

The purpose of this experiment is to examine the force of interaction between two charged, small metal-coated balls, and examine how it varies with r, the distance between the centers of the balls.

Since you will only be able to determine the positions of the charged prod, x_{prod}, and hanging charged ball, x_{ball}, from the frames of the movie, you will need to use the laws of mechanics to determine the relationship between the Coulomb force on the hanging charged ball and the distance, x_{ball}, that the ball is displaced as the thread is displaced through an angle θ. Thus, you should be able to calculate the Coulomb force on q_1 (on the hanging charged ball) as a function of the distance $r = x_{ball} - x_{prod}$ between q_1 and q_2.

Before proceeding with the video analysis, you will need to determine how the force the prod exerts on the hanging ball (the Coulomb force, $\vec{F}_e$) depends on the position of the hanging ball. This will allow you to calculate the values of the Coulomb force using the data from your video analysis. The situation is shown in the diagram below.

1. Draw a vector diagram with arrows showing the direction of each of the forces on the charged hanging ball with charge q_1 and mass m, including the gravitational force, $\vec{F}_g$, the tension in the thread, $\vec{F}_T$, and a horizontal electrostatic force, $\vec{F}_e$, due to the charge on the prod.

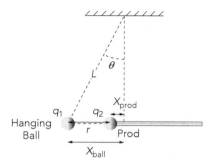

2. Show that, if there is no motion in the vertical direction, then

$$\sum F_y = 0$$

and, therefore, $F_T = \dfrac{ma_g}{\cos\theta}$

where a_g is the gravitational acceleration.

3. Show that, if there is no motion along the horizontal direction, then

$$\sum F_x = F_e - F_T \sin\theta = 0$$

4. Show that $F_e = ma_g \tan\theta$.

5. Find $\tan\theta$ as a function of x_{ball} and L. **Hint:** Start by finding y as a function of x_{ball} and L, in the diagram below.

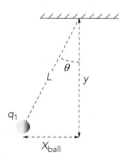

6. Combine the equations you found in (4) and (5) to find the magnitude of the electrostatic force, F_e, as a function of L, x_{ball}, m, and a_g.

7. Finally, use the assumption that θ is a small angle (that is, the displacement of the charged ball, x_{ball}, is small compared to the length, L, of the thread from which it is suspended) to simplify your equation.

Now, you can turn to the task of measuring how the charged ball moves as a function of its horizontal distance from the prod, r. Then, using the equation you just obtained, you can analyze the video data to determine F_e as a function of r.

You will need the following:

- *RealTime Physics Electricity and Magnetism* experiment configuration files
- computer-based video analysis software

Activity 2-3: Analyzing the Digital Video to Find Distances and Forces

1. Open the experiment file **Coulomb's Law (L01A2-3).** This will open a movie in the video analysis software.

2. Record the mass of the hanging ball and the vertical distance, L, from the point of suspension to the center of the ball. This information is given in the movie, usually on the first frame.

   m:_____   L:_____

3. Use the video analysis software to analyze the video frames by recording: (a) the position of the suspended charged ball and (b) the position of the prod ball in each frame of the video.

4. Your measurements need to be converted to meters. If your video analysis software is not set up to do this automatically, figure out how to do it.

5. Set up calculated columns in the video analysis software or a spreadsheet to calculate the Coulomb force F_e from your data for x_{ball}, and also to calculate r, the distance between the suspended ball and the prod ball from your data for x_{ball} and x_{prod}.

Activity 2-4: Analyzing the Data to Examine the Relationship of Coulomb Force to r

1. If your video analysis software has the capability, display the graph of F_e as a function of r. (If not, open the experiment file **Force vs. r (L01A2-4).** This will open a table and axes. Enter your data to plot a graph of F_e as a function of r.)

2. Use the **modeling feature** of the software to examine the mathematical relationship between F_e and r.

3. Affix any graphs to these sheets.

Question 2-3: Can you fit a plot of F_e vs. r with an equation of the form $F = C/r^2$ where C is a constant?

Question 2-4: Does the $F = C/r^2$ relationship seem to hold? Explain.

Question 2-5: Describe the most plausible sources of uncertainty in your data.

INVESTIGATION 3: THE ELECTRIC FIELD

Most of the forces you have studied up until now resulted from the direct action or contact of one object on another. (The only exception was the gravitational force.) From your observations in Investigations 1 and 2, it should be obvious that charged objects can also exert forces on each other at a distance. How can that be? The action at a distance that characterizes electrical forces is in some ways inconceivable to us. How can one charge feel the presence of another and detect its motion with only empty space in between? Since all atoms and molecules are thought to contain electrical charges, physicists currently believe that all contact forces are really electrical forces involving small separations. So, even though forces acting at a distance seem inconceivable to most people, physicists believe that *all* forces act at a distance.

To describe action at a distance, Michael Faraday introduced the notion of an *electric field* emanating from a collection of charges and extending out into space. More formally, the electric field due to a known collection of charges is represented by a vector at every point in space. Thus, the electric field vector, $\vec{E}$ is defined as the force, $\vec{F}$, that would be experienced by a very small positive charge (called a *test* charge) at a point in space, divided by the magnitude of the charge q_o. Thus, the electric field is in the direction of the force $\vec{F}$ on the small positive test charge and has a magnitude of

$$\vec{E} = \vec{F}/q_o$$

To carry out some qualitative measurements of the electric field in simple situations, you will need

- roll of Scotch Magic© tape
- glass or acrylic rod and polyester or silk cloth
- hard black plastic, Teflon, or rubber rod and fur

Activity 3-1: Electric Field Vectors from a Positively Charged Rod

To investigate the vector nature of an electric field, you can use a piece of Scotch Magic© tape with a positive charge on it as the *test* charge.

1. Charge a piece of tape (about 10 cm long) positively. To recall how to do this, refer back to Investigation 1.

2. Charge up the glass or acrylic rod (positively), and hold it pointing vertically. Assume that the charge on the tip of the glass rod is the source of the electric field.

3. Now hold the test charge (tape) vertically, and move it around the rod. Note the direction and magnitude of the force at various locations and different distances around the rod. **Note:** The charged tape is really not a point charge, but it can still give you an idea of the direction of the electric field in the vicinity of a point.

> **Comment:** By convention physicists always place the tail of the E-field vector at the point in space of interest rather than at the charge that causes the field.

Question 3-1: What is the direction of the electric field at various points around the rod?

Question 3-2: How does the magnitude of the electric field vary with distance from the tip of the rod?

Question 3-3: Do a qualitative sketch of some electric field vectors around the tip of the rod at points marked on the circles in the diagram below. The length of the vector should roughly indicate the *relative* magnitude of the field, and of course, the direction of the vector should indicate the direction of the field. Don't forget to put the tail of the vector at the location of interest, not at the location of the glass rod.

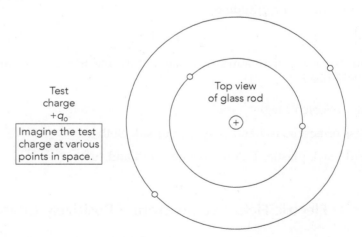

Test charge $+q_0$

Imagine the test charge at various points in space.

Top view of glass rod

$\oplus$

Extension 3-2: Electric Field from a Negatively Charged Rod

Use the hard plastic or Teflon rod to create an electric field resulting from a negative charge distribution.

Question E3-4: What is the direction and relative magnitude of the electric field around the rod?

Question E3-5: Sketch the electric field vectors with both *magnitude* and *direction* drawn in on the basis of the force experienced by the test charge at the points on the circles in the diagram below.

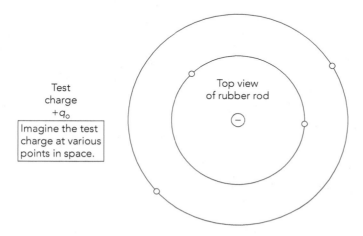

Name_____ Date_____ Partners_____

HOMEWORK FOR LAB 1
ELECTRIC CHARGES, FORCES AND FIELDS

1. You have two charged pieces of Scotch Magic© tape. How would you determine if they have like or unlike charges? What would you need to determine if they are charged positively or negatively?

2. Two like charges are separated by some distance. Describe quantitatively what will happen to the force exerted by one charge on the other if

    a. The distance between the charges is doubled

    b. The distance between the charges is halved

    c. One of the charges is replaced by a charge of the same magnitude but opposite sign

3. Charge q_2 is 2.5×10^{-9} C and charge q_1 has mass 0.20 g. The separation r is 5.0 cm, and the angle θ is 15 degree. Find q_1 (magnitude and sign).

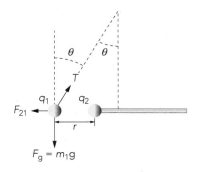

4. Find the magnitude and direction of the electric field at the position of q_1 produced by q_2.

5. Draw electric field vectors at the points A, B, C, D, and E caused by the two point charges shown. (**Hint:** The field at any point is the vector sum of the fields from each of the charges.)

C

B

o   A   D   o   E
+q   +q

6. Find the magnitude of the force on the −3.0 C charge, and show its direction on the diagram. (The distance scale is 5.0 cm per division.)

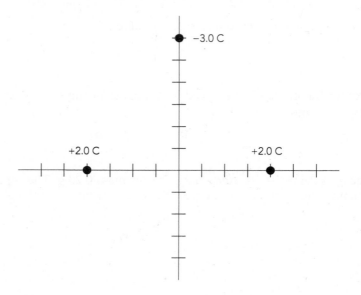

PRE-LAB PREPARATION SHEET FOR
LAB 2—ELECTRIC FIELDS, FLUX AND GAUSS' LAW

(Due at the beginning of lab)

Directions:
Read over Lab 2 and then answer the following questions about the procedures.

1. In Activity 1-1, how will you be able to sketch the electric field lines for simple charge distributions?

2. Describe the model you will use in Activity 1-3 to examine the dependence of flux on angle.

3. How will you analyze the data in Activity 1-3?

4. In Activity 1-4, what will you be comparing the net number of electric field lines coming out of a closed surface to?

5. In Activity 2-2, do you predict that the excess charge on the can will reside on the inside surface of the can or the outside surface? Briefly describe the observations you will make to test your prediction.

LAB 2:
ELECTRIC FIELDS, FLUX AND GAUSS' LAW

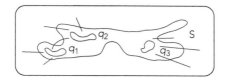

> *. . . before Maxwell people considered physical reality*
> *. . . as material points . . . After Maxwell they considered physical reality as*
> *continuous fields. . .*
>
> — A. Einstein

OBJECTIVES

- To understand how electric field lines can be used to describe the magnitude and direction of an electric field in a small region of space.

- To discover how the electric flux passing through a small area is related to the magnitude of the electric field, as well as the area and its orientation relative to the direction of the electric field.

- To understand why $\Sigma E \cos \theta \, \Delta A$ over a closed surface is proportional to the number of lines of flux passing through the surface.

- To understand the relationship between the flux passing through a closed surface and the charge enclosed by that surface (Gauss' law) for a two-dimensional situation.

- To explore how Gauss' law predicts where excess charge can be found on a conductor.

OVERVIEW

In Lab 1 you explored how Coulomb's law can be used to calculate the force on a test charge due to charges in its vicinity. You have learned that electric field is defined as a vector that describes the direction and magnitude of the force exerted on a positive test charge *per unit charge*. Although the electric field can be calculated using Coulomb's law, it is very difficult when many charges are present at different locations. However, it is possible to calculate the electric field using a completely different formulation of Coulomb's law. This formulation is known as Gauss' law, and it involves relating the electric field surrounding a collection of charges to the net amount of charge inside a closed surface. The Gauss' law formulation is a very powerful tool for calculating electric fields in situations where the distribution of charge is very *symmetric*. Gauss' law can be proven to be mathematically equivalent to Coulomb's law. In this lab you are going to take an approach to Gauss' law that is not so mathematical.

Before exploring Gauss' law, however, you will see how to represent the electric field in a region of space by using *electric field lines*. Next you will define electric flux, a quantity that is related to the electric field lines passing through a surface. You will try to discover Gauss' law by drawing closed "surfaces" around various charges or groups of charges and seeing how many electric field lines pass in and out of the surface.

Reality Check: Always remember that the electric field, and the various schemes for representing it in space, are just aids to representing electrostatic forces mathematically. The force is the only real physical thing that can be measured! Gauss' law can seem confusing at times, but it has given us a much better understanding of interactions between charged particles.

INVESTIGATION 1: ELECTRIC FIELD LINES, ELECTRIC FLUX AND GAUSS' LAW

Electric Field Lines

In Lab 1, you represented the electric field produced by a configuration of electric charges by arrows with magnitude and direction at each point in space. This is the *conventional representation* of a "vector field." An alternative representation of the vector field involves the use of *electric field lines*. Unlike an electric field vector that is an arrow located at a point in space with magnitude and direction, electric field lines are continuous. Instead of defining electric field lines formally, you can discover what they represent by using a computer simulation.

You will need:

- field line simulation program (or printouts of electric field lines generated using the simulation)
- computer

Activity 1-1: Simulation of Electric Field Lines from Point Charges

1. Open the field line simulation program. (Or use a printout of the electric field lines from such a program.)

2. Set a single +1 (arbitrary unit) charge somewhere on the computer screen and run the program. After a few minutes sketch the field lines in the space below, on the left. Then place a −1 charge somewhere on the screen, and sketch the field lines associated with it on the right. Show the direction of the electric field on each line by placing an arrow on it.

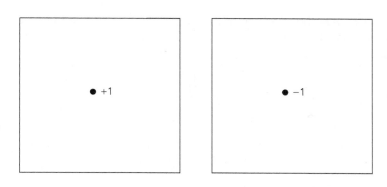

Question 1-1: How many lines are there in the drawings? Are the lines more dense (closely spaced) or less dense near the charge? Explain. How do the directions of the lines depend on the sign of the charge?

3. Try another magnitude of charge. You don't need to sketch the result.

Question 1-2: What is the magnitude of your new charge? How many lines are shown in the simulation? How do the number of lines compare to the number in Question 1-1?

Question 1-3: If you know the magnitude of the charges, describe the rule for telling how many lines will come out of or into a charge in this simulation. Explain your answer based on your observations.

4. Repeat the exercise using two charges of the *same magnitude* but opposite sign. After the simulated lines are drawn, sketch them in the space below. Indicate how much charge you used on the diagram.

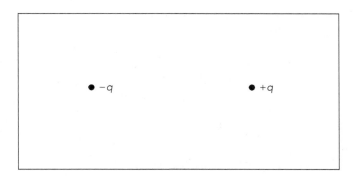

Question 1-4: Are the lines beginning or ending on a charge more dense near it or far away from it? How does the direction of the lines depend on the sign of the charge?

Question 1-5: Summarize the properties of electric field lines. What does the number of lines signify? What does the direction of a line at each point in space represent? What does the density of the lines represent?

Defining Electric Flux

We can think of an electric charge as having a number of electric field lines diverging from it or converging on it in such a way that the number of lines is proportional to the magnitude of the charge. Now we can explore the mathematics of enclosing charges within surfaces and seeing how many electric field lines pass through the surface. *Electric flux* is defined as a measure of the number of electric field lines passing through a surface. In defining "flux" we are constructing a mental model of lines streaming out from the surface area surrounding each unit of charge like streams of water or rays of light. Physicists do not really think of charges as having anything real streaming out from them, but the mathematics that best describes the forces between charges is the same as the mathematics that describes streams of water or rays of light. Let's explore the behavior of this mathematical model.

It should be obvious that the number of field lines passing through a surface depends on how that surface is oriented relative to the electric field lines. The orientation of a small surface of area A can be defined by a normal vector that is perpendicular to the surface and has a magnitude equal to the surface area, as shown in the figure below. By convention, the normal vector points away from the *outside* of the surface. The normal vectors are pictured for small surfaces of area A that make up the ends of a cylindrical box. In the picture the inside of each surface is light gray and the outside is dark gray.

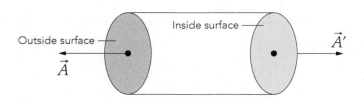

Activity 1-2: Drawing Normal Vectors

Use the definition of normal to an area given above to draw normal "area" vectors to the surfaces shown below. Label each vector as $\vec{A}$. Let the length of the normal vector in cm be equal in magnitude to each area in cm^2.

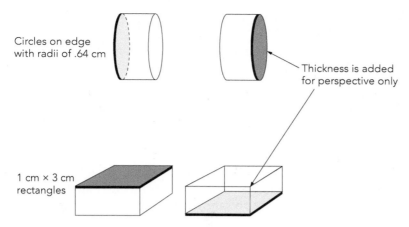

Circles on edge with radii of .64 cm

Thickness is added for perspective only

1 cm × 3 cm rectangles

By convention, if an electric field line passes from the inside to the outside of a closed surface we say the flux is positive. If the field line passes from the outside to the inside of a surface, the flux is—by definition—negative.

Prediction 1-1: How does the flux through a surface depend on the angle between the normal vector to the surface and the electric field lines? For example, what is the electric flux when the angle is 90 degrees? What happens to the electric flux as you rotate the surface at various angles between 0 degrees (or 0 radians) and 180 degrees (or π radians) with respect to the electric field vectors? What do you think?

In order to test your predictions in a quantitative way, you can use a mechanical model of some electric field lines and of a surface. The model is made with a 10 by 10 array of nails poking up at 1/4" intervals through a piece of foil covered foam insulation. The "surface" is a wire loop painted white on the "outside." You will need

- 100 nails (approximately 4" long) to represent field lines
- 5" × 5" square of Styrofoam
- 5" × 5" square of ¼" graph paper affixed to the Styrofoam to make an evenly spaced grid for the nails
- square wire loop (4" × 4") to represent a "surface"
- protractor
- data logging software
- *RealTime Physics Electricity and Magnetism* experiment configuration files

The finished mechanical model for flux is shown below.

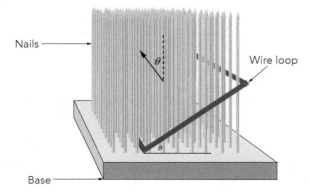

Nails

θ

Wire loop

θ

Base

Activity 1-3: Flux as a Function of Angle

1. Use your mechanical model, a protractor and some calculations to fill in the data table below. **Note:** The angle, θ, is between the normal to the area (surface of loop) and the direction of the field lines (nails). If the normal is pointing generally in the same direction as the field, the angle is between 0 and 90°. If the normal is pointing generally opposite to the field, the angle is more than 90°, and the flux is negative.

2. Open the experiment file **Flux vs. Angle (L02A1-3),** and enter your data for flux vs. angle.

3. Find the mathematical relationship between flux and angle. Guess the mathematical relationship (there are a few hints in this lab), and try to confirm your guess by plotting a new graph of flux as a function of angle and comparing it to the graph of your data. (Create a new **calculated column.** Use the function of angle that you want to test, and adjust the parameters until you get the best agreement with the graph of the actual data. Try different functions until you have found the one that describes the functional relationship between your data and the angle. (Use the angle in radians for trigonometric functions.)

4. Alternatively, you can do a manual fit to the trigonometric function $A \sin (B \theta + C) + D$, and find the parameters A, B, C, and D.

Φ (flux = number of lines)	θ (deg) (normal pointing up)	Φ (flux = number of lines)	θ (deg) (normal pointing down)
100		−10	
90		−20	
80		−30	
70		−40	
60		−50	
50		−60	
40		−70	
30		−80	
20		−90	
10		−100	
0			

Print all of your graphs and affix to these sheets.

Question 1-6: What is the mathematical relationship between flux and angle? Explain based on your observations.

A Mathematical Representation of Flux through a Surface

By definition, the relationship between flux, Φ, and angle, θ, for a uniform electric field $\vec{E}$ is $\Phi = E\,A\cos\theta$, where θ is the angle between the normal to the surface and the electric field vector, $\vec{E}$. Flux is a scalar quantity.

If the electric field is not uniform or if the angle between the surface and electric field varies from point to point on the surface, then flux needs to be calculated by breaking the surface into infinitely many infinitesimally small areas (so that $\Delta\Phi = E\cos\theta\,\Delta A$) and then adding the fluxes through all the areas. This gives

$$\Phi = \sum\Delta\Phi = \sum E\,\Delta A\cos\theta \qquad \text{[flux through a surface]}$$

Some surfaces like that of a sphere or that representing a rectangular box are closed surfaces as they have no holes or breaks in them.

In order to study the amount of flux passing through closed surfaces, you will need

- field line simulation program (or field line printouts from one)
- computer

Activity 1-4: Discovering Gauss' Law in Flatland

Since it is charge that produces the electric field, the electric flux passing through a closed surface must depend on the enclosed charge. But what is the dependence? Suppose you lived in a two-dimensional world in which all charges and electric field lines were constrained to lie in a flat, two-dimensional space. Of course, mathematicians call such a space a plane. You can open the field line simulation program again and set the program to sketch lines for some nutty creative mix of charges (or use already printed ones). Don't be too creative or the lines will take forever to sketch out! You should do the following:

1. Open up the simulation, place some positive *and* negative charges at different places on the screen and start the program to calculate and display the electric field lines in two dimensions.

2. In the space below, sketch your computer screen showing the configuration of the charges and associated "e-field" lines, or include a screen shot of the computer screen.

3. Draw arrows on each of the lines indicating in which direction a *small* positive test charge would move along each line. **Note:** "small" means that the test charge does not exert large enough forces on the charges that create the e-field to cause the field to change noticeably when it is brought near them.

Having done all this preparation, you should be ready to discover how the net number of lines enclosed by a surface is related to the net charge enclosed by the surface.

4. Draw three two-dimensional closed "surfaces" in pencil on your diagram, above. Some of them should enclose charge, and some should avoid enclosing charge.

5. Draw at least one more closed "surface" that encloses charges, but zero *net* charge.

You are ready to discover how the net number of lines enclosed by a surface is related to the net charge enclosed by the surface.

6. Count the *net* lines of flux coming out of (leaving) each "surface." (**Note:** The *net* number of lines is defined as the number of lines coming out of the surface minus the number of lines going in.)

7. Fill in the table below for three different Flatland "surfaces."

	Charge enclosed by the arbitrary surface			Lines of flux in and out of the surface		
	$+q$	$-q$	q_{net}	Lines$_{out}$	Lines$_{in}$	Lines$_{net}$
1						
2						
3						
4						

Question 1-7: What is the apparent relationship between the net flux (net number of lines) passing through an imaginary surface and the net charge enclosed by the two-dimensional "surface"? Explain, based on your simulated observations.

Gauss' Law in Three Dimensions

If you were to repeat the simulated exploration you just performed in a three-dimensional space, what do you think would be the appropriate expression for Gauss' law? You will explore this in the next activity.

Activity 1-5: Statements of Gauss' Law

Question 1-8: Express the three-dimensional form of Gauss' law in words. **Hints:** What would be the three-dimensional equivalent of the closed curves you drew in steps (4) and (5)? Does it still make sense in three-dimensions to talk about the net number of lines coming out of a closed surface?

Question 1-9: Using the results from Activities 1-3 and 1-4, and your answer to Question 1-8, express Gauss' law using an equation. (**Hint:** The net flux should be on one side of the equation, and the other side should have an expression for the net charge enclosed by the surface. Don't worry about the actual value of the proportionality constant.

INVESTIGATION 2: USING GAUSS' LAW

An electrical conductor has some of its electrical charges free to move about (i.e., not bound to atoms). If a free charge in a conductor experiences an electric field, it will move under the influence of that field. *Thus,* we can conclude that if there are no charges moving within a conductor, the electric field within the conductor is zero.

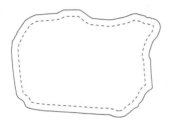

Let's consider the conductor shown above that has been touched by a charged black plastic, Teflon, or rubber rod so that it has an excess of negative charge on it. Where does this charge go if it is free to move about? Is it distributed uniformly throughout the conductor? If we know that $\vec{E} = 0$ everywhere inside a conductor, we can use Gauss' law to figure out where the excess charge on a conductor is located. This is the subject of the next two activities.

Activity 2-1: Where is the Excess Charge in a Conductor?

Question 2-1: Consider the conductor in the diagram above, with an excess charge Q. As just argued above, there is zero electric field within the conductor. What is the amount of excess charge enclosed by the dashed Gaussian surface drawn just below the surface of the conductor? **Hint:** Use Gauss' law, and the fact that $\vec{E} = 0$ everywhere inside the conductor.

Question 2-2: If the conductor has excess charge and it can't be within the Gaussian surface according to Gauss' law, then what's the only place this excess charge can be?

Question 2-3: Forgetting for the moment about Gauss's law but recalling that like charges repel each other, is the conclusion you reached in answering Question 2-2 reasonable physically? Explain. **Hint:** How can charges that are repelling each other get as far apart as possible from each other?

Since you used Gauss's law to predict that excess charge in a conductor would move to the outside surface of the conductor, let's check this prediction experimentally. We can do this by using the following:

- soup can
- Styrofoam pad to insulate the can from the table
- black plastic, Teflon, or rubber rod
- fur
- aluminum-covered Styrofoam ball hanging from a light string

The test of Gauss's law that you are about to perform is referred to as the Faraday Ice Pail experiment, although some people attribute it to Benjamin Franklin.

Activity 2-2: The Faraday Ice Pail Experiment

1. Place the can on top of the Styrofoam pad.
2. Rub the rod vigorously with the fur for several seconds. Next rub the rod on the *outside* of the can several times to transfer charge to it. Repeat this entire procedure at least five times to transfer a significant charge.
3. Bring the aluminum-covered ball hanging from the string close to *but not touching* the outside of the can.

Question 2-4: What happens when the ball is brought close to the can?

4. Now touch the ball to the outside of the can, and then move it a very short distance away.

Question 2-5: What happens after the ball is touched to the side of the can and then moved away? Explain the difference in behavior of the ball in (3) and (4).

Question 2-6: What do you conclude about the presence of charge on the outside of the can? Explain based on your observations.

5. Charge the can again as in (2). *Be sure to transfer as much charge as possible from the rod to the can.*

6. Next lower the uncharged aluminum-covered ball down the middle of the can until it is close to the center of the can. Then move it close to *but not touching* the inside of the can.

Question 2-7: What happens when the ball is brought close to the inside of the can?

7. Now touch the ball to the inside of the can and move it slightly away, inside the can.

Question 2-8: What happens after the ball is touched to the inside of the can and then moved away?

Question 2-9: What do you conclude about the presence of charge on the inside of the can?

8. This time transfer charge to the *inside* of the can by charging the rod as in (2), but rubbing the rod on the *inside* of the can. *Be sure to transfer as much charge as possible from the rod to the can.*

9. Repeat observations (6) and (7).

Question 2-10: What do you conclude about the presence of charge on the *inside* of the can? Explain.

10. Charge the inside of the can again as in (8).

11. Repeat observations (3) and (4).

Question 2-11: What seems to have happened to the charge you put on the *inside* of the can? Explain based on your observations in (9) and (11).

Question 2-12: Describe how Gauss's law explains your observations.

HOMEWORK FOR LAB 2:
ELECTRIC FIELDS, FLUX AND GAUSS' LAW

1. Describe the properties of electric field lines.

   a. If the total number of lines leaving or converging on a charge doubles, what does that tell you about the magnitude of the charge?

   b. What does the direction of a line at each point in space represent?

   c. What does the density of lines represent?

2. Based on your observations of electric field lines in this lab, sketch several electric field lines originating from the +q charge. Then sketch how many lines would originate from a charge of +3q. (Note that your sketches don't need to be exactly correct, but they should illustrate that you understand the properties of electric field lines discussed in Question 1.)

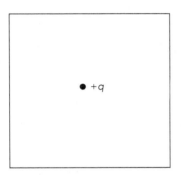
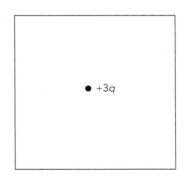

3. For situations a through e, use the definition of flux that you developed in Activity 1-3 to determine the electric flux passing through each surface. The electric field has magnitude 5.0 N/C at the location of the surface of area of 1.5 cm^2. Show your calculations.

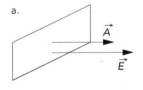
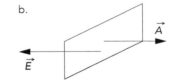
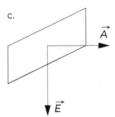

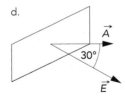
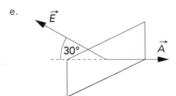

4. What is the net electric flux through the closed surface in each case shown below? Assume that 5 lines leave a charge of +q or terminate on a charge of −q. (Assume that all of the surfaces are three-dimensional.) Use the net number of field lines leaving the surface as a measure of flux. Explain in the spaces below how you arrived at your answers.

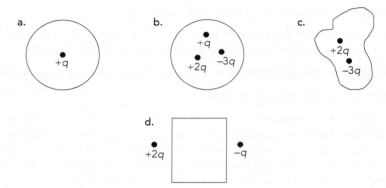

a.

b.

c.

d.

5. How does Gauss' law explain that excess charge on a conductor must reside on the outside surface?

Pre-Lab Preparation Sheet for
Lab 3—Electrical and Gravitational Potential

(Due at the beginning of lab)

Directions:
Read over Lab 3 and then answer the following questions about the procedures.

1. In Activity 1-3, how do you think that the work done along path ac will compare to that along path adc ?

2. What will you use in Activity 1-3 to measure force?

3. In Activity 2-2, what types of points on the conducting paper will you be trying to find?

4. In Activity 2-2, what device will you use to measure potential differences?

5. Once you have sketched equipotentials in Activity 2-2, what procedures will you use to sketch the electric field lines?

LAB 3:
ELECTRICAL AND GRAVITATIONAL POTENTIAL

> *Electricity seems destined to play a most important part in the arts and industries. The question of its economical application to some purposes is still unsettled, but experiment has already proved that it will propel a street car better than a gas jet and give more light than a horse.*
>
> —Ambrose Bierce (1842–1913)

OBJECTIVES

- To understand the similarities in the mathematics used to describe gravitational and electrical forces.

- To review the mathematical definition of work and potential energy in a conservative force field as the sum of the product of displacement and the force component in the direction of the displacement:

$$W = \Sigma\, F \cos \theta\, \Delta s$$

- To understand the definition of electrical potential, or *voltage*.

- To map and examine the potential distribution in two dimensions for several simple sets of electrodes.

- To learn the relationship between electric field lines and equipotential surfaces.

OVERVIEW

The enterprise of physics is ultimately concerned with identifying and mathematically describing the fundamental forces of nature. Nature offers us several fundamental forces, which include a strong force that holds the nuclei of atoms together, a weak force that helps us describe certain kinds of radioactive decay in the nucleus, the force of gravity, and the electromagnetic force.

Two kinds of force dominate our everyday reality—the gravitational force acting between masses and the Coulomb force acting between electrical charges. The gravitational force allows us to describe how objects near the Earth's surface are attracted toward it, and how the moon revolves around the Earth and planets revolve around the Sun. The genius of Newton was to realize that objects as diverse as falling apples and revolving planets are all moving under the action of the same gravitational force.

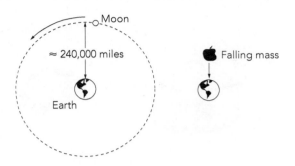

Similarly, the Coulomb force allows us to describe how one charge "falls" toward another or how an electron orbits a proton in a hydrogen atom.

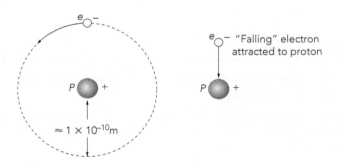

The fact that both the Coulomb and the gravitational forces lead to objects falling and to objects orbiting around each other suggests that these forces might have the same mathematical form.

In this lab we will explore the mathematical symmetry between electrical and gravitational forces for two reasons. First, it's beautiful to behold the unity that nature offers us as we use the same type of mathematics to predict the motion of planets and galaxies, the falling of objects, the flow of electrons in circuits, and the nature of hydrogen and other atoms. Second, what you already learned about the influence of the gravitational force on a mass and the concept of gravitational potential energy can aid your understanding of the forces between charged particles.

We will introduce the concept of electrical potential energy in analogy to that of gravitational potential energy. An understanding of *electrical potential difference,* commonly called *voltage,* is essential to understanding electrical circuits. The lab will culminate with the measurement of electrical potential differences due to electric fields produced by simple conductors.

Let's begin our discussion with the familiar expression for the Coulomb force exerted on charge 2 by charge 1, $\vec{F}_{1\rightarrow2}$,

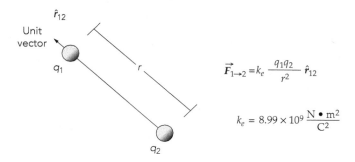

$$\vec{F}_{1\rightarrow2} = k_e \frac{q_1 q_2}{r^2} \hat{r}_{12}$$

$$k_e = 8.99 \times 10^9 \frac{\text{N} \cdot \text{m}^2}{\text{C}^2}$$

where k_e is a constant that equals 9.0×10^9 N m^2/C^2. The force $\vec{F}_{2\rightarrow1}$ of charge 2 on charge 1 is equal in magnitude and opposite in direction to $\vec{F}_{1\rightarrow2}$.

Coulomb did his experimental investigations of this force in the eighteenth century by exploring the forces between two small charged spheres. You have already examined the Coulomb force law in Lab 1.

Newton's discovery of the universal law of gravitation came the other way around. He thought about orbits first. This was back in the seventeenth century, long before Coulomb began his studies. A statement of Newton's universal law of gravitation describing the force experienced by mass 2 due to the presence of mass 1 is

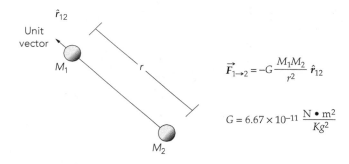

$$\vec{F}_{1\rightarrow2} = -G \frac{M_1 M_2}{r^2} \hat{r}_{12}$$

$$G = 6.67 \times 10^{-11} \frac{\text{N} \cdot \text{m}^2}{\text{Kg}^2}$$

where G is equal to 6.67×10^{-11} Nm^2/Kg^2. Just as with the electrostatic force, $\vec{F}_{2\rightarrow1}$ is equal in magnitude and opposite in direction to $\vec{F}_{1\rightarrow2}$.

Electrical and gravitational forces are very similar, since essentially the same mathematics can be used to describe orbital and linear motions due to either electrical or gravitational interactions of the tiniest fundamental particles or the largest galaxies.

Activity 1-1: The Electrical vs. the Gravitational Force

Examine the mathematical expression for the two force laws.

Question 1-1: What is the same about the two force laws?

Question 1-2: What is different about the force laws? For example, is the force between two like masses attractive or repulsive? How about two like charges? What part of each equation determines whether like charges or masses are attractive or repulsive?

Question 1-3: If there were negative mass would two negative masses attract or repel? How would a negative mass interact with a positive one? Do you think negative mass could exist?

Which Force Is Stronger—Electrical or Gravitational?

Let's peek into the hydrogen atom and compare the gravitational force on the electron due to interaction of its mass with that of the proton to the electrical force between the two particles as a result of their charges. In order to do the calculation, you'll need to use some well-known values:

Electron: $m_e = 9.1 \times 10^{-31}$ Kg $\quad q_e = -1.6 \times 10^{-19}$ C

Proton: $\quad m_p = 1.7 \times 10^{-27}$ Kg $\quad q_p = +1.6 \times 10^{-19}$ C

Distance between the electron and proton: $r = 1.0 \times 10^{-10}$ m

Activity 1-2: The Electrical vs. the Gravitational Force in the Hydrogen Atom

1. Calculate the magnitude of the electrical force the proton exerts on the electron in a Hydrogen atom. Is the force attractive or repulsive?

2. Calculate the magnitude of the gravitational force the proton exerts on the electron in a hydrogen atom. Is the force attractive or repulsive?

Question 1-4: Which force is larger? By what factor? Calculate the ratio of the larger to the smaller.

Question 1-5: Which force are you more aware of on a daily basis? If your answer does not agree with that in Question 1-4, explain why.

Work in a Gravitational Field – A Review

Let's review some old definitions in preparation for tackling the ideas of work and energy expended in moving through an electric field. Work is a scalar rather than a vector quantity defined mathematically as:

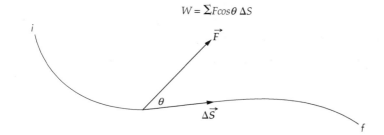

$$W = \Sigma F \cos\theta \, \Delta S$$

The sum is over a path between the initial (i) and final (f) locations, where each element of $F \cos\theta \, \Delta s$ represents the projection of F along Δs at each place along the path.

Let's review the procedures for calculating work and for examining whether the work done in a conservative force field is independent of path. You will measure the work done moving a cart along two different paths, ac and adc, as shown in the diagram below.

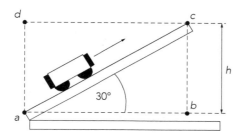

In order to take measurements you will need the following equipment:

- inclined track
- low friction cart
- computer-based laboratory system
- force sensor
- *RealTime Physics Electricity and Magnetism* experiment configuration files
- meterstick
- protractor

Activity 1-3: Measuring and Calculating Earthly Work

1. Set the inclined track to an angle of about 30° above the horizontal. Measure the length of the track and the elevation of its end.

   L = _____   h = _____

2. Open experiment configuration file **Measuring Force (L03A1-3)** to use the force sensor to measure the force you are exerting.

3. Use the force sensor to apply the force that is just necessary to move the cart at a constant speed up along the inclined track. Be sure to apply the force parallel to the surface of the track.

4. Use the **analysis feature** in the software to determine the average force.

    Average force = _____

5. Use the definition of work to calculate the work you performed. Show your calculations.

    $$W = \underline{\hspace{3cm}}$$

Prediction1-1: Suppose that instead you raise the cart directly to the same height as the top of the inclined track along path ad at a constant speed, again using the force sensor to measure the applied force. Then you move the cart horizontally from point d to point c, while it is suspended from the force sensor. How will the work done in this case compare to the work done pulling the cart along the track?

6. Move the cart in the way described in Prediction 1-1, measuring the average force for both parts of the motion.

    Average force along ad _____ along dc _____

7. Again, use the *definition of work* to calculate the work done in moving the cart along path adc. Show your calculations.

    $$W = \underline{\hspace{3cm}}$$

Question 1-6: How does the work calculated in (5) compare to that calculated in (7). Does this agree with your prediction? Why or why not?

Conservative Forces

It takes work to lift an object under the influence of the Earth's gravitational force. This increases the gravitational potential energy of the object. Lowering the object releases the gravitational potential energy that was stored when it was lifted. When you studied the gravitational force, you applied the term *conservative* to it because it allows the recovery of *all* of the stored energy. You have now found experimentally that the work required to move an object under the influence of the gravitational force is path independent. This is an important property of any *conservative* force.

Given the mathematical similarity between the Coulomb force and gravitational force laws, it should come as no surprise that experiments confirm that the Coulomb force is also *conservative*. Again, this means that the work needed to move a charge is independent of the path taken between the points. And from the definition of work

$$W = \sum_{A}^{B} F \cos \theta \Delta s = \sum_{A}^{B} q E \cos \theta \Delta s$$

is the work done by the electric field E to move a small test charge q between points A and B.

Activity 1-4: Work Done on a Charge Traveling in a Uniform Electric Field

1. Suppose that a small positive test charge q is moved a distance d from point A to point B along a path that is parallel to a uniform electric field of magnitude E.

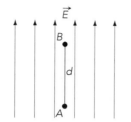

Question 1-7: What is the work done by the field on the charge? Show your calculation.

Question 1-8: How does the form of this equation compare to the work done on a mass m traveling a distance d parallel to the almost-uniform gravitational force near the surface of the Earth?

2. The charge q is now moved a distance d from point A to point B in a uniform electric field of magnitude E, but this time the path is perpendicular to the field lines.

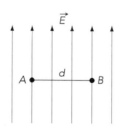

Question 1-9: What is the work done by the field on the charge? Show your calculation.

3. The charge q is now moved a distance d from point A to point B in a uniform electric field of magnitude E. The path lies at a 45° angle to the field lines.

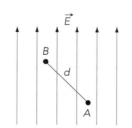

Question 1-10: What is the work done by the field on the charge? Show your calculation.

Electrostatic Potential Energy and Potential

The change in electrostatic potential energy is defined as the negative of the work done by the electrostatic force in moving a charge from point A to point B using any path consisting of a series of small increments of length Δs. Thus, using U as the symbol for potential energy,

$$\Delta U^{elec} = U_B^{elec} - U_A^{elec} = -\sum_A^B qE \cos \theta \Delta s$$

The electric potential difference $\Delta V = V_B - V_A$ is defined as the change in electrical potential energy, ΔU^{elec}, *per unit charge*.

Comment: The potential difference has units of joules per coulomb. Since a J/C is defined as *one volt*, the potential difference is often referred to as *voltage*.

Activity 1-5: The Equation for Potential Difference

Using the equation above for potential energy change, and the definition of potential difference, write the equation for the potential difference as a function of E and Δs.

Note: The simplest charge configuration is a single point charge. As you saw in Lab 1, a positive point charge q produces an electric field that points radially outward in all directions. The potential difference between any two points in space A and B can be found using calculus. The result is that if A is infinitely far away, and B is a finite distance r from a point charge q, then the potential difference, V, is given by the expression

$$V = k_e(q/r)$$

In this case, the reference point for the potential difference is at infinity, and the potential difference is simply referred to as "the potential."

INVESTIGATION 2 : EQUIPOTENTIAL SURFACES

Sometimes it is possible to move along a surface in a non-zero electric field without doing any work. Thus, it is possible to remain at the same potential anywhere along such a surface. If an electric charge can travel along a surface without any work being done, the surface is defined as an *equipotential surface*.

Consider the three different charge configurations shown below. Where are the equipotential surfaces? What shapes do they have?

Activity 2-1: Electric Field Lines and Equipotentials

Suppose you have a positive test charge and you move it in space some distance from the charge below. (The arrows represent electric field lines.)

Question 2-1: Given that the electric field is non-zero, what path could you move a test charge along without doing any work, i.e., for which $E \cos \theta \Delta s$ is always zero? What is the shape of the equipotential surface? (Remember that in general you can move in three dimensions.)

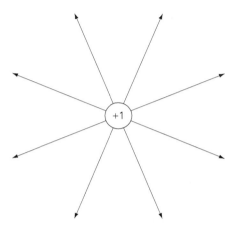

Question 2-2: What is the relationship between the direction of the equipotential lines you have drawn (representing that part of the equipotential surface that lies in the plane of the paper) and the direction of the electric field lines? Explain.

Prediction 2-1: Draw some equipotential surfaces for the charge configuration shown below that results from two charged metal plates placed parallel to each other. (The field lines drawn represent a uniform electric field, which is correct in the central region of the plates, away from the edges.) What is their shape?

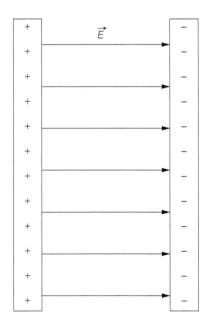

Prediction 2-2: Draw some equipotential surfaces for the electric dipole charge configuration shown below. (Again, the lines represent the electric field, and the arrows represent its direction.) What is their shape?

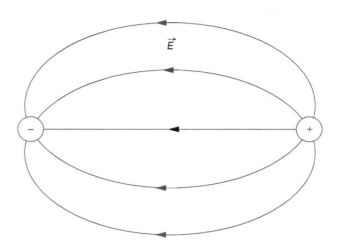

The purpose of the following activity is to explore the pattern of potential differences in the space around two different charge distributions. This pattern of potential differences can be related to the electric field caused by the charges that lie on the conducting surfaces.

You will be using pieces of carbonized paper with conductors painted on them in different shapes. When a battery is used to create a potential difference between the two conductors, this can *simulate* the pattern of equipotential lines associated with metal electrodes in air. (**Warning:** This is only a simulation of "reality!") One of the papers has two small point-like circular conductors painted on it and the other has two linear conductors–as shown in the diagrams on the next two pages.

There are several activities associated with this laboratory project. First, you will use a battery to set up a voltage across a paper with circular (point) electrodes or line electrodes. You can then use a digital voltmeter to trace several equipotential lines for the electrode configuration you choose. This should give you a feel for the measurements and, hopefully, confirm your ideas of what the pattern of equipotentials are for the two-dimensional situation you have chosen. You will need:

- equipotential plotting sheets made from carbonized paper (with line and circular "point" electrodes painted on with conducting paint)

- digital voltmeter

- 6 V battery or power supply

- 2 alligator clip leads

- 2 point sensors with leads

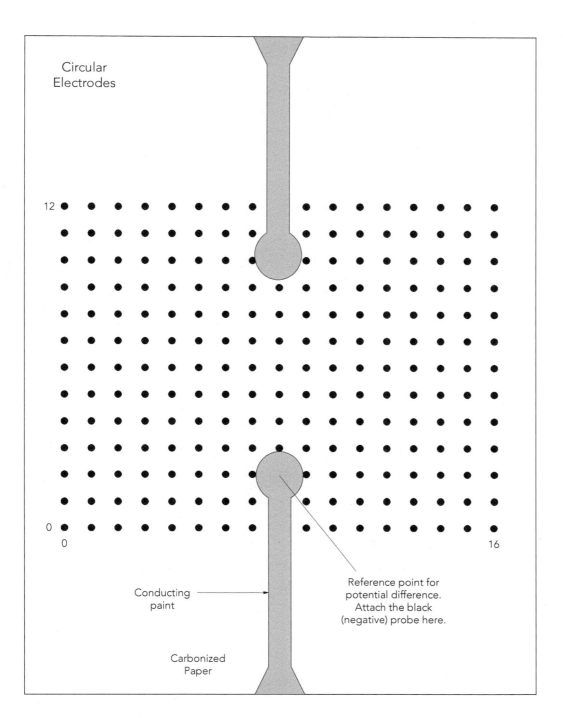

Circular
Electrodes

12

0

0   16

Conducting
paint

Carbonized
Paper

Reference point for
potential difference.
Attach the black
(negative) probe here.

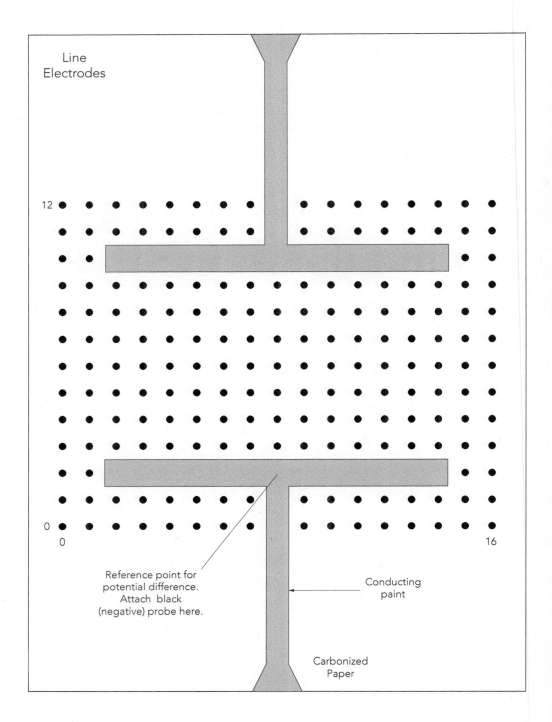

Line Electrodes

12

0

0   16

Reference point for
potential difference.
Attach black
(negative) probe here.

Conducting
paint

Carbonized
Paper

Activity 2-2: Mapping Equipotentials and Field Lines from "Point" Electrodes on Conducting Paper

Carry out these procedures for the circular, "point" electrodes.

1. Set up the circuit as shown below with the "point" electrodes. Use the alligator clips to connect the terminals of the battery to each of the electrodes on the paper.

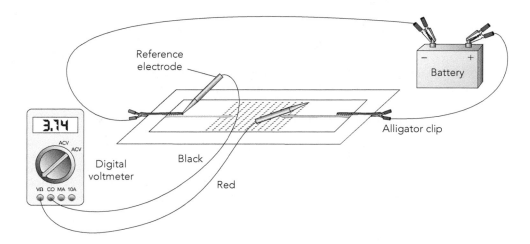

2. Turn on the digital multimeter to one of the DC voltage ranges. Place the tip of the black probe (plugged into the COM input) from the multimeter on the electrode where the negative terminal of the battery is connected.

3. First place the tip of the red probe (plugged into the V-Ω-A input) on the same electrode as the black probe.

Question 2-3: What is the reading when both probes touch the same electrode? Why?

4. Now, place the tip of the red probe on any other location on the paper. The reading on the multimeter will display the potential difference between the two points. Examine what happens if you reverse the probes.

Question 2-4: What happens when you reverse the probes? Why?

5. With the black probe back in place on the reference electrode, poke around with the red probe to find points that are at the same potential. Trace out equipotential lines for potential differences of +1V, +2V, and +3V. Sketch the location of the lines to scale on the diagram.

Question 2-5: Are the equipotentials what you expected? Did they agree with your predictions? Explain.

6. From your knowledge that electric field lines are always perpendicular to equipotential lines, sketch some field lines on your diagram in a different color. Be sure to put arrows on them to indicate the direction of the electric field.

Question 2-6: Explain how you knew how to draw your field lines. Do these field lines agree with your predictions?

Extension 2-3: Mapping Equipotentials and Field Lines from Parallel Plate Electrodes on Conducting Paper

1. Repeat (1)–(5) above for the line electrodes.
2. Repeat (6) above to draw field lines for the line electrodes.

Question E2-7: Describe the similarities and differences between the equipotentials for the "point" electrodes and those for the line electrodes.

Question E2-8: Describe the similarities and differences between the field lines for the "point" electrodes and those for the line electrodes.

HOMEWORK FOR LAB 3
ELECTRICAL AND GRAVITATIONAL POTENTIAL

1. In each of the drawings below, $E = 4.0 \, \text{N/C}$ and $d = 1.5$ m. For each diagram calculate:

   a. The work done by the electric field in moving a $+5.0 \times 10^{-9}$ C charge from A to B.

   b. The difference between the charge's potential energy at B and at A.

   c. The potential difference between A and B.

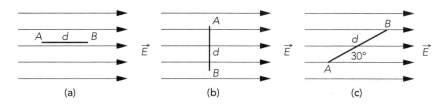

   (a)   (b)   (c)

2. What is the relationship between equipotential surfaces (or lines in two dimensions) and the direction of electric field lines?

3. How much work is done moving a charge along an equipotential surface (line)? Explain.

4. Sketch at least five equipotential lines for the electric field lines shown below. Add arrows to the field lines to show the direction of the field. Explain how you decided where to draw the lines.

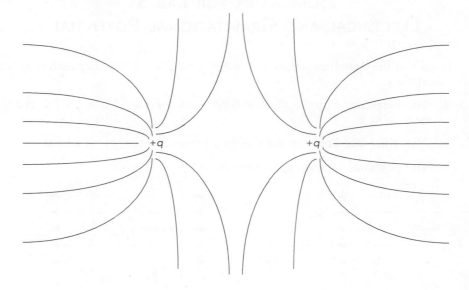

5. Explain how you would use a voltmeter to find equipotential lines for electrodes painted on carbonized conducting paper with conducting paint.

6. How would you draw in the electric field lines once you located the equipotential lines on the carbonized conducting paper?

PRE-LAB PREPARATION SHEET FOR
LAB 4—BATTERIES, BULBS, AND CURRENT

(Due at the beginning of lab)

Directions:
Read over Lab 4 and then answer the following questions about the procedures.

1. Explain why in Activity 1-1 the angle irons will be charged in several different ways.

2. Sketch below one arrangement of the battery, bulb, and wire (other than the one in Figure 4-3) that you will study in Activity 1-2.

3. At this time, which of the models shown in Figure 4-6 for how current flows in a circuit do you favor? Why?

4. Describe briefly how you will test the models for current. What device will you use to measure current?

5. Sketch the symbols for a battery and for a light bulb below.

LAB 4:
BATTERIES, BULBS, AND CURRENT*

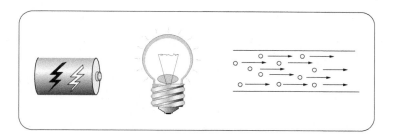

*You cannot teach a man anything; you
can only help him to find it within himself.*

—Galileo

OBJECTIVES

- To understand how a potential difference (voltage) can cause an electric current to flow in a conductor.

- To learn to design and construct simple circuits using batteries, bulbs, wires, and switches.

- To learn how to draw circuit diagrams using standard circuit symbols.

- To understand the measurement of current and voltage using microcomputer-based sensors.

- To understand what currents will flow in various parts of simple DC circuits, and why.

OVERVIEW

In the following labs, you are going to discover and extend theories about electric charge and potential difference (voltage), and apply them to electric circuits. What you learn will be one of the most practical parts of the whole physics course, since electric circuits form the backbone of modern technology. Without an understanding of electric circuits we wouldn't have lights, air conditioners, automobiles, telephones, TV sets, dishwashers, computers, or photocopying machines.

A *battery* is a device that generates an electric potential difference (voltage) from other forms of energy. The type of batteries you will use in these labs are known as chemical batteries because they convert internal chemical energy (energy stored in chemical bonds) into electrical energy.

As a result of a potential difference, electric charge is repelled from one terminal of the battery and attracted to the other. However, no charge can flow out of a battery unless there is a conducting material connected between its terminals.

---

*Some activities in this lab have been adapted from those designed by the Physics Education Group at the University of Washington, included in Workshop Physics, Module 4.

If this conductor happens to be the filament in a small light bulb, the flow of charge will cause the light bulb to glow.

In this lab you are going to explore how charge flows in wires and bulbs when energy is provided by a battery. You will be asked to develop and explain some models that predict how the charge flows. You will also be asked to devise ways to test your models using computer-based current and voltage sensors that measure the rate of flow of electric charge (current) and the potential difference (voltage), respectively, and display these quantities on a computer screen using data collection and analysis software.

INVESTIGATION 1: MODELS THAT DESCRIBE CURRENT FLOW

What is electric current? As you have seen in Lab 1, the forces between objects that are rubbed in particular ways can be attributed to a property of matter known as *charge* (static electricity). Most textbooks assert that the electric currents through the wires connected to a battery are charges in motion. How do we know this? Perhaps current is something else—another phenomenon.

This question received a great deal of attention from Michael Faraday, a famous early-nineteenth-century scientist. Faraday studied the effects of electricity from animals including electric eels summarized his results in a table like the one shown in Figure 4-1, and concluded that "electricity, whatever may be its source, is identical in its nature."[1]

	Physiological effect	Magnetic deflection	Magnets made	Spark	Heating power	True chemical action	Attraction and repulsion	Discharge by hot air
1. Voltaic electricity	X	X	X	X	X	X	X	X
2. Common electricity	X	X	X	X	X	X	X	X
3. Magneto-electricity	X	X	X	X	X	X	X	
4. Thermo-electricity	X	X	+	+	+	+		
5. Animal electricity	X	X	X	+	+	X		

Figure 4-1: Faraday's table. The X's denote results obtained by Faraday and the +'s denote positive results found later by other investigators.

The purpose of the first activity is to compare carriers of the current produced by a battery to the static charges deposited by rubbing materials together. You will observe a demonstration using the following materials:

- 2 metal angle irons (approx. 15 cm long)
- foil-covered Styrofoam ball on a string (2.5 cm diameter)
- 300-V battery pack or power supply

---

[1]Faraday, M. "Identity of Electrocutes Derived from Different Sources," in *Experimental Researches in Electricity, Vol. I*, Taylor and Francis, London. (Reprinted by Dover Publications, New York, 1965, p. 76).

- hard rubber (plastic or Teflon©) rod and fur

- glass or acrylic rod and polyester, felt or silk cloth

- electroscope

- alligator clip leads

- Wimshurst or Van de Graaff generator (optional)

Activity 1-1: Comparing Stuff from a Battery to the Rubbing Stuff

1. The angle irons, ball, and electroscope will be set up as shown in Figure 4-2.

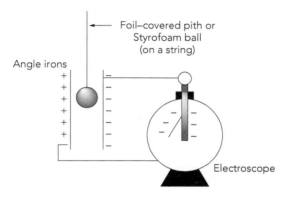

Figure 4-2: Apparatus for detecting charge.

2. The metal angle irons will be charged in two or more of the following ways:

    a. *Electrostatic Charging by Rubbing.* Stroke one plate with a rubber rod that has been rubbed with the fur. Repeat this several times. Stroke the other plate with the glass rod that has been rubbed with the polyester cloth.

    b. *Charging with a Battery.* Connect a wire from the negative terminal of the power supply to one of the angle irons. At the same time connect a wire from the positive terminal to the other plate.

    c. *Charging with a Wimshurst or Van de Graaff Generator.* Connect a wire from one of the two terminals of the generator to one angle iron, and a wire from the other terminal to the other angle iron.

3. Observe whether the different charging methods have different effects on the electroscope and on the ball when it is dangled between the two metal angle irons.

4. The metal angle irons will then be separated so the gap between them is just barely bigger than the diameter of the foil covered ball, and the ball will be carefully placed between them. You should observe something unusual.

Question 1-1: What happened to the electroscope when the angle irons were charged using the familiar rubbing method described in (a) above? Why?

Question 1-2: What happened when the ball was placed between the angle irons? Use your understanding of the attraction and repulsion of different types of charges to explain why this unusual phenomenon happened.

Question 1-3: Describe what happened when the power supply charged the angle irons. What differences (if any) did you observe in the response of the electroscope and the ball to the charges on the angle irons?

Question 1-4: If the Wimshurst or Van de Graaff generator was also used, describe what happened. Did you observe any difference in the response of the ball to the charges on the angle irons? How about the response of the electroscope?

Question 1-5: Do the charges generated by rubbing and those from the power supply cause different effects? If so, describe them. Do the charges generated in these two ways seem different?

The rate of electric charge flow is called *electric current*. If charge Δq flows through the cross section of a conductor in time Δt, then the average current can be expressed mathematically by the relationship

$$\langle i \rangle = \frac{\Delta q}{\Delta t}$$

Instantaneous current is defined as the charge per unit time passing through a particular part of a circuit at an instant in time. It is defined by using the limit:

$$i = \text{limit } \frac{\Delta q}{\Delta t}$$

as $\Delta t \rightarrow 0$

The unit of current, called the ampere (A), represents the flow of one coulomb of charge in a time of one second. Another common unit is the milliampere (mA) (1 ampere = 1000 milliamperes). Usually people just refer to current as "amps" or "milliamps."

In the next activity, you can begin exploring electric current by lighting a bulb with a battery. You will need the following:

- flashlight bulb (#14)
- flashlight battery (1.5-V D cell, alkaline and very fresh)
- wire (6 inches or more in length)

Activity 1-2: Arrangements that Cause a Bulb to Light

Use the materials listed above to find some arrangements in which the bulb lights and some in which it does not light. For instance, try the arrangement shown in Figure 4-3.

Figure 4-3: A wiring configuration that might cause a battery to light a bulb.

Question 1-6: Sketch two arrangements for which the bulb lights.

Question 1-7: Sketch two arrangements for which the bulb *doesn't* light.

Question 1-8: Describe as fully as possible the conditions needed for the bulb to light. Explain why the bulb fails to light in the arrangements drawn in answer to Question 1-7.

Next you will explore materials connected between a battery and a bulb allow the bulb to light. Since it seems that something flows from the battery to the bulb, we refer to materials that allow this flow as *conductors* and those that don't as *nonconductors or insulators*.
You will need

- common objects (paper clips, pencils, coins, rubber bands, fingers, paper, keys, etc.)

Activity 1-3: Other Materials Between the Battery and Bulb

Set up the single wire, battery, and bulb so that the bulb lights, e.g., one of the arrangements drawn in your answer to Question 1-6. Then, with the help of your partner, stick a variety of the common objects available between the battery and the bulb.

Question 1-9: List some materials that allow the bulb to light.

Question 1-10: List some materials that prevent the bulb from lighting.

Question 1-11: What types of materials seem to be conductors? What types seem to be nonconductors?

Having trouble holding your circuits together? Let's make it easier by using a battery holder and a bulb socket. While we're at it, let's also add a switch in the circuit. In addition to the materials you've already used, you will need:

- battery holder (for a D cell)
- several wires (6 inches or more in length)
- flashlight bulb socket
- contact switch

Activity 1-4: Using a Battery Holder, Bulb Socket, and Switch

1. Examine the bulb socket carefully. Observe what happens when you unscrew the bulb.

2. Examine the bulb closely. Use a magnifying glass, if available. Figure 4-4 shows the parts of the bulb that are hidden from view.

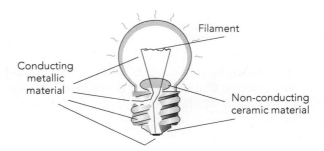

Figure 4-4: Wiring inside a light bulb.

Question 1-12: Why is the filament of the bulb connected as shown in Figure 4-4?

Question 1-13: Explain how the bulb socket works. Why doesn't the bulb light when it is unscrewed?

Prediction 1-1: If you wire up the configuration shown in Figure 4-5, will the bulb light with the switch open (i.e., so no contact between the wires is made)? Closed (i.e., so that contact is made)? Neither time? Explain your predictions.

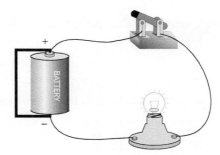

Figure 4-5: A circuit with a battery, switch, and bulb holder.

3. Wire the circuit shown in Figure 4-5 and test it.

4. Leave the switch closed so that the bulb remains on for 10–20 seconds. Feel the bulb.

Question 1-14: What did you feel? Besides giving off light, what happens to the bulb when there is a current flowing through it?

Question 1-15: What do you conclude about the path needed by the current to make the filament heat up and the bulb glow? Explain based on all the observations you have made so far.

You are now ready to explore models for what happens to current in a circuit. The circuit to be considered is the one shown in Figure 4-5 when the switch is closed. Figure 4-6 shows some common models for current flow in this circuit.

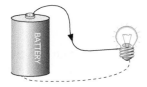

Model A
There will be no electric current left to flow in the bottom wire since the current is used up lighting the bulb.

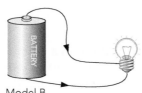

Model B
The electric current will flow in a direction toward the bulb in both wires.

Model C
The direction of the current will be in the direction shown, but there will be less current flowing in the return wire since some of the current is used up lighting the bulb.

Model D
The direction of the current flow will be as shown, and the magnitude will be the same in both wires.

Figure 4-6: Four alternative models for current in a simple circuit.

Prediction 1-2: Which of the models do you think best describes how current flows in the circuit? After you've made your own prediction, explain your reasoning to your partner(s).

After you have discussed various ideas with your partner(s) and chosen your favorite model, test your predictions by using one or more current sensors in your circuit to measure current. In addition to the battery, bulb, and wires you used above, you will need:

- computer-based laboratory system
- two current sensors
- *RealTime Physics Electricity and Magnetism* experiment configuration files

The current sensor is a device that measures current and displays it as a function of time on the computer screen. It will allow you to explore the current flow at different locations and under different conditions in your electric circuits.

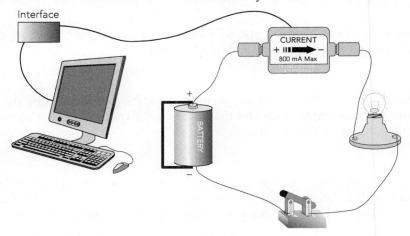

Figure 4-7: A circuit with a battery, bulb, switch, and current sensor connected to the computer interface.

To measure the current through a part of the circuit, you must break open the circuit at the point where you want to measure the current, and insert the current sensor. *That is, disconnect the circuit, put in the current sensor, and reconnect with it in place.* For example, to measure the current flowing in the top wire of the circuit in Figure 4-5, the current sensor should be connected as shown in Figure 4-7.

Note that the current sensor measures both the *magnitude* and *direction* of the current flow. A current that flows in through the + terminal and out through the − terminal (in the direction of the arrow) will be displayed as a positive current. Thus, if the current measured by the sensor is positive, you know that the current must be clockwise in Figure 4-7 from the + terminal of the battery, through the bulb, through the switch, and toward the − terminal of the battery.

On the other hand, if the sensor measures a negative current, then the current must be counterclockwise in Figure 4-7 (flowing into the − terminal and out of the + terminal of the sensor).

Figure 4-8a shows a simplified diagram representing a current sensor connected as shown in Figure 4-7.

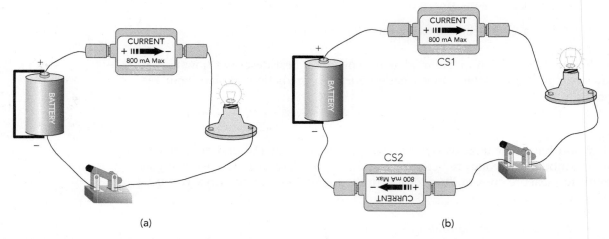

(a)   (b)

Figure 4-8 (a) Current sensor connected to measure the current out of the + terminal of the battery and into the bulb. (b) Two current sensors, one connected as in (a) and the other connected to measure the current out of the switch and into the − terminal of the battery.

Look at Figure 4-8b and convince yourself that if the currents measured by current sensors 1 and 2 are both positive, this shows that the current is in a clockwise direction around all parts of the circuit.

Design measurements that will allow you to choose the model (or models) that best describe the actual current through the circuit. (For example, to see if the current has a different magnitude or direction at different points in the circuit [model B or model C in Figure 4-6], you should connect two current sensors in various locations around the circuit as in Figure 4-8b, to measure the current.)

Prediction 1-3:   Use Table 4-1 to describe how the signs and magnitudes of the readings of current sensor 1 and current sensor 2 in the circuit in Figure 4-8b would compare with each other for each of the current models described in Figure 4-6.

Table 4-1

	Sensor	Positive, negative, or zero?	CS1 > CS2, CS1 < CS2, or CS1 = CS2?
Model A	CS1		
	CS2		
Model B	CS1		
	CS2		
Model C	CS1		
	CS2		
Model D	CS1		
	CS2		

Activity 1-5: Developing a Model for Current in a Circuit

1.   Be sure that current sensors 1 and 2 are plugged into the interface.

2.   Open the experiment file called **Current Model (L04A1-5).** The two sets of axes that follow should appear on the screen. The top axes display the current through sensor 1 and the bottom the current through sensor 2.

 The amount of current through each sensor is also displayed digitally on the screen.

3.   Be sure to **calibrate** the sensors, or **load the calibration. Zero** the sensors befor they are connected in the circuit.

4.   To begin, set up the circuit in Figure 4-8b. **Begin graphing,** and try closing the switch for a couple of seconds and then opening it for a couple of seconds. Repeat this several times during the time when you are graphing. Sketch your graphs on the axes that follow, or **print** your graphs and affix them over the axes.

Note: You should observe carefully whether the current flowing through both sensors is essentially the same or if there is a *significant* difference (more than a few percent).

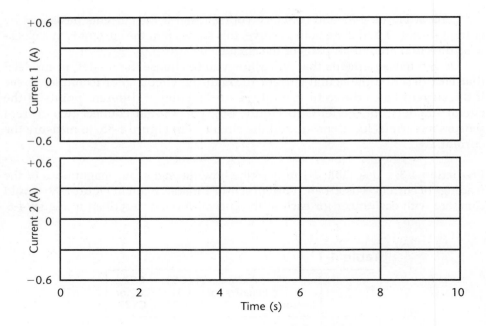

Question 1-16: How can you determine if an observed difference in the currents read by sensor 1 and sensor 2 is real or if it is the result of small calibration differences and normal fluctuations in the sensor readings?

Question 1-17: Did you observe a *significant* difference in the currents flowing at these two locations in the circuit, or was the current the same?

5. Try any other tests you need to decide which current model among the choices in Figure 4-6, or any others you come up with, seems most reasonable. Sketch circuit diagrams for the arrangements you used, showing where the sensors were connected. **Print** all graphs, label them, and affix them below.

Question 1-18: Based on your observations, which model seems to correctly describe the behavior of the current in the circuit in Figure 4-5. Explain carefully based on your observations.

INVESTIGATION 2: CURRENT AND POTENTIAL DIFFERENCE

Now that you have been wiring circuits and drawing diagrams of them you may be getting tired of drawing pictures of the batteries, bulbs, and switches. There are a series of symbols that have been created to represent circuits. A few electric circuit symbols are shown in Figure 4-9.

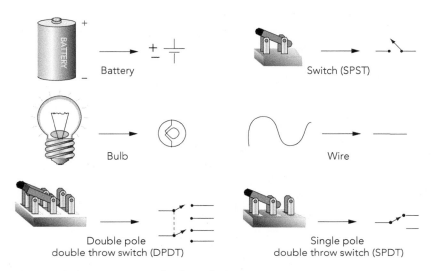

Figure 4-9: Some common circuit symbols.

Using these symbols, the circuit from Figure 4-5 with a switch, bulb, wires, and battery can be sketched as on the right in Figure 4-10.

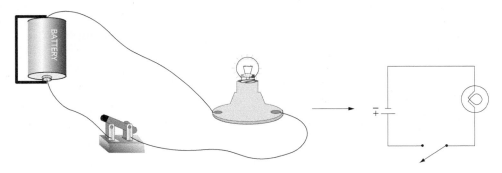

Figure 4-10: A circuit sketch and corresponding circuit diagram.

Activity 2-1: Drawing Circuit Diagrams

Sketch a nice, neat "textbook" style circuit diagram for each of the circuits shown below on the left

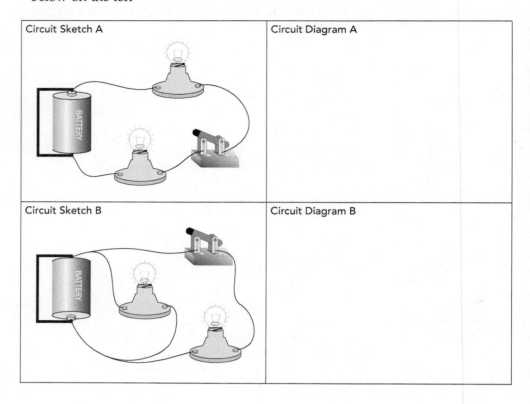

| Circuit Sketch A | Circuit Diagram A |
| Circuit Sketch B | Circuit Diagram B |

Question 2-1: On the battery symbol, which line represents the positive terminal—the long one or the short one? [**Note:** You should try to remember this convention for the battery polarity because some circuit elements, such as *diodes*, behave differently if the battery is turned around so it has opposite polarity.]

Activity 2-2: Inventing and Constructing Electric Circuits

Since you know how to get a bulb to light, you can get more practice with wiring by designing some simple electrical devices and building some circuits. Then you can do a more careful exploration of the behavior of currents and voltage in basic DC circuits.

For your design projects, you can use some of the following equipment:

- 3 #14 bulbs and sockets
- 1.5-V D cell very fresh, alkaline battery with holder
- 6 alligator clip leads
- single-pole–single-throw (SPST) switch
- single-pole–double-throw (SPDT) switch
- double-pole–double-throw (DPDT) switch

Invent and construct one of the electric circuits described below. Sketch the circuit in the space below.

1. *Christmas Tree Lights:* Suppose you want to light up your Christmas tree with three bulbs. What happens if one of the bulbs fails? (Don't break the bulb! You can simulate failure by loosening a bulb in its socket.) Figure out a way to connect all three bulbs so that the other two will still be lit if any one of the bulbs burns out.

2. *Lighting a Tunnel:* The bulbs and switches must be arranged so that a person walking through a tunnel can turn on a lamp for the first part of the tunnel and then turn on a second lamp for the second half of the tunnel in such a way that the first one is turned off.

3. *Entry and Exit Light Switches:* A room has two doors. Light switches at both doors are wired so that either switch turns the lights in the room on and off.

Circuit diagram for circuit #____

Question 2-2: Describe how your circuit works, and why you connected it in the way you did.

Extension 2-3: Inventing Other Circuits

Invent and construct one or both of the other electric circuits described in Activity 2-2, or invent your own circuit that performs a certain task. Sketch each circuit diagram and describe how the circuit works in the space below.

There are actually two important quantities to consider in describing the operation of simple DC electric circuits. One is *current*, and the other is *potential difference*, often referred to as *voltage*. In Activity 1-5 you measured the current flow at two different positions in a circuit. Now, as an introduction to our studies of more complex circuits, let's actually measure *both current and voltage* in a familiar circuit.

In addition to the equipment you have been using so far, you will need:

- computer-based laboratory system
- two voltage sensors
- two current sensors
- *RealTime Physics Electricity and Magnetism* experiment configuration files

Figure 4-11 shows the symbols we will use to indicate a current sensor or a voltage sensor.

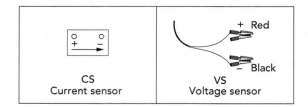

Figure 4-11: Symbols for current sensor and voltage sensor.

Figure 4-12a shows our simple circuit from Figure 4-5 with voltage sensors connected to measure the voltage *across* the battery and the voltage *across* the bulb. The circuit is drawn again symbolically in Figure 4-12b. Note that the word *across* is very descriptive of how the voltage sensors are connected.

Activity 2-4: Measuring Potential Difference (Voltage) and Current

1. To set up the voltage sensors, first unplug the current sensors from the interface and plug in the voltage sensors.

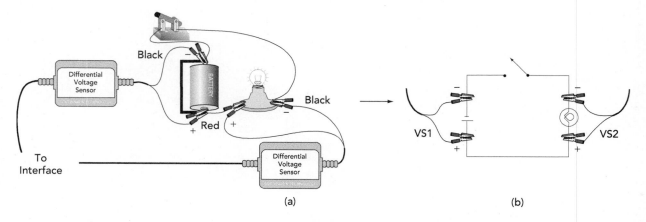

(a) (b)

Figure 4-12: Two voltage sensors connected to measure the voltages across the battery and the bulb.

2. Open the experiment file called **Two Voltages (L04A2-4a)** to display the axes for two voltage sensors that follow.

3. **Zero** the voltage sensors with them disconnected from the circuit and both clips of each attached together.

4. Connect the circuit shown in Figure 4-12, but *do not connect the sensors yet*.

5. First connect both the + and the − clips of one voltage sensor to the *same point* in the circuit. Close the switch.

6. Finally connect the + clip to the + end of the battery and the − clip to the side of the bulb indicated with a + in Figure 4-12. Close the switch.

Question 2-3: What do you conclude about the voltage when the voltage sensor leads are connected to the same point or to the two ends of the same wire?

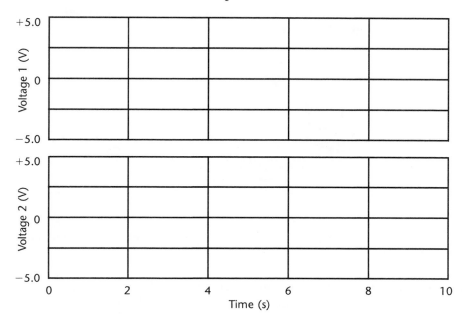

Prediction 2-1: In the circuit in Figure 4-12, how do you expect the voltage across the battery to compare to the voltage across the bulb when the switch is open? Closed? Explain.

7. Now test your prediction by connecting the voltage sensors to measure the voltage *across* the battery and the voltage *across* the bulb simultaneously.

8. **Begin graphing,** and close and open the switch several times. **Print** your graphs and affix them over the axes above, or sketch them on the axes.

Question 2-4: What do you conclude about the voltage across the battery and the voltage across the bulb when the switch is open and when it is closed? Are the graphs the same? Why or why not? What is going on as the switch is closed and opened?

9. Now connect a voltage and a current sensor so that you are measuring the voltage *across* the battery and the current *through* the battery at the same time (see Figure 4-13).

10. Open the experiment file called **Current and Voltage (L04A2-4b)** to display the current and voltage axes that follow.

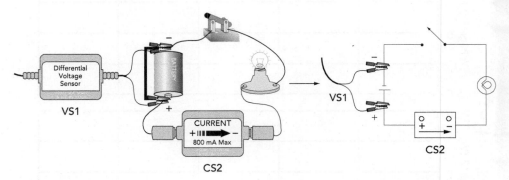

Figure 4-13: Sensors connected to measure the voltage across the battery and the current through it.

11. **Begin graphing,** and close and open the switch several times, as before. Sketch your graphs, or **print** them and affix over the axes.

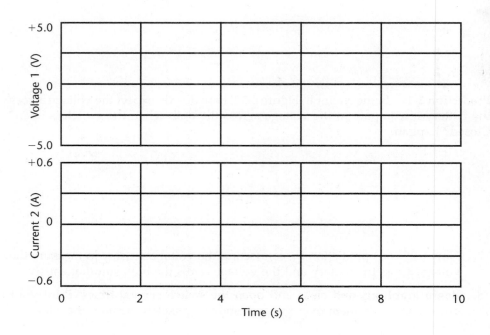

Question 2-5: Explain the appearance of your current and voltage graphs. What happens to the current through the battery as the switch is closed and opened? What happens to the voltage across the battery?

12. Find the voltage across and the current through the battery when the switch is closed and the bulb is lit. (You can use the digital display on the computer screen.)

   Average voltage: ____   Average current: ____

Prediction 2-2: Now suppose you connect a second bulb in the circuit, as shown in Figure 4-14. How do you think the voltage across the battery will compare to that with only one bulb? Will it change significantly? Explain.

13. Connect the circuit with two bulbs and test your prediction. Again measure the voltage across and the current through the battery with the switch closed.

   Average voltage: ____   Average current: ____

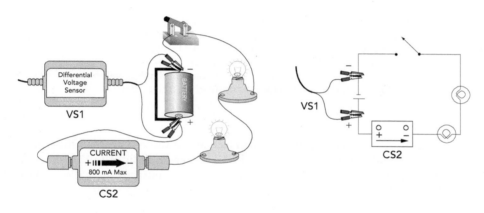

Figure 4-14: Two bulbs connected with voltage and current sensors.

Question 2-6: When you added the second bulb to the circuit, did the current through the battery change significantly (i.e., by more than a few percent)?

The current did not change through this process.

Question 2-7: When you added the second bulb to the circuit did the voltage across the battery change significantly (i.e., by more than a few percent)?

The voltage did not change.

Question 2-8: Does the battery appear to be a source of constant current, constant voltage, or neither when different elements are added to a circuit?

No, the battery is not a source of constant current.

INVESTIGATION 3: AN ANALOGY TO CURRENT AND RESISTANCE

You found in Activity 1-5 that current is not used up in flowing through a bulb, but this may seem counterintuitive to you. Also, how can we explain that there is less current in the circuit with two bulbs instead of one? Lots of physics teachers have invented analogies to help explain these electric circuit concepts. One approach is to construct a model of a gravitational system that is in some ways analogous to the electrical system you are studying. This is shown in Figure 4-5.

It is believed that the electrons flowing through a conductor have frequent collisions that slow them down and change their directions. Between collisions each electron accelerates and finally staggers through the material with an average drift velocity, $\langle \vec{v}_{\text{drift}} \rangle$.

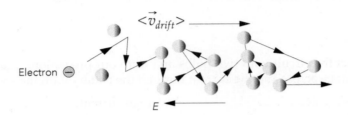

Figure 4-15: A simplified depiction of an electron in a uniform electric field staggering through a conductor as a result of collisions. Despite the constant force to the right caused by the electric field, these collisions cause the electron to move through the conductor with a constant average velocity, $\langle \vec{v}_{\text{drift}} \rangle$, the "drift velocity" (instead of accelerating).

As you saw in Investigation 2, we can talk about the *resistance* to flow of electrons that materials offer. A wire has a low resistance. A light bulb has a much higher resistance. Special electric elements that resist current are called *resistors*. You will examine the behavior of these in electric circuits in future labs.

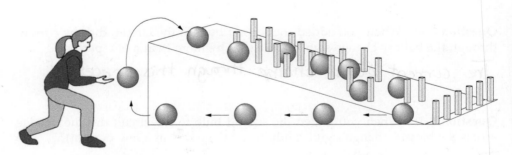

Figure 4-16: An analog to electric current and resistance.

It is possible to use a two-dimensional mechanical analog to model this picture of current through conductors. You should note that the real flow of electrons is a three-dimensional affair. The diagram for the two-dimensional analog is reproduced in Figure 4-16.

Extension 3-1: Drawing an Analogy

Look at the current and resistance analog in Figure 4-16, or view the movie in experiment file **Current and Resistance Analog (L04E3-1)** and then answer the following questions.

Question E3-1: What part of the picture represents the action of the battery? What represents the electric charge and current? What part represents the collisions of electrons? Explain.

Question E3-2: What ultimately happens to the "energy" given to the balls by the "battery"? What plays the role of the bulb? How is this energy loss exhibited in the circuit you wired that consists of a battery, two wires, and a bulb?

Question E3-3: How does this model help explain the fact that electric current doesn't decrease as it passes through the bulb?

Question E3-4: How does this model help explain the fact that electrons move with a constant average speed v_{drift}, rather than having a constant acceleration caused by the constant electric field?

Question E3-5: In this model what would happen to the "ball" current if the drift velocity doubled? What can you do to the ramp to increase the drift velocity?

HOMEWORK FOR LAB 4
BATTERIES, BULBS, AND CURRENT

1. Is there any difference between the static charges generated by rubbing rods with fur or silk and the charges that flow (from a battery) through wires in an electric circuit? Give evidence for your answer.

2. For the circuit on the right, indicate whether the statements below are TRUE or FALSE. If a statement is TRUE, briefly describe the evidence from this lab which supports this statement. If a statement is FALSE, give a correct statement, and briefly describe the evidence from this lab which supports this new statement.

   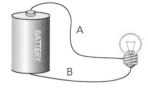

   a. The current flows from the battery, through wire A, through the bulb, and then back to the battery through wire B.

   b. Since current is used up by the bulb, the current in wire B is smaller than the current in wire A.

   c. The current flows toward the bulb in both wires A and B.

   d. If wire B is disconnected, but wire A is left connected, the bulb will still light, but if wire A is disconnected and wire B is left connected, the bulb will not light.

   e. A current sensor will read the same magnitude if connected to measure the current in wire A, and then connected to measure the current in wire B.

3. Name the circuit element represented by each of the following symbols:

   a.   b.   c.   d.

4. Use the symbols shown in Question 3 to draw at least one of the circuits you worked on in Activity 2-2.

5. Draw below a circuit diagram for the circuit in Question 2 with one current sensor hooked up to measure the current in wire A and a voltage sensor hooked up to measure the voltage across the light bulb. Also include a switch in the circuit to turn the bulb on and off. Use correct symbols for all circuit elements.

6. Consider the two circuits below. All bulbs and all batteries are identical. Compare the voltage across the battery in the left circuit to that in the right circuit. *Describe the evidence in this lab for your answer.*

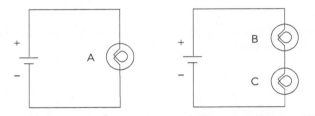

PRE-LAB PREPARATION SHEET FOR
LAB 5—CURRENT IN SIMPLE DC CIRCUITS

(Due at beginning of lab)

Directions:
Read over Lab 5 and then answer the following questions about the procedures.

1. What do you predict for the rankings of the brightness of bulbs A, B, and C in Figure 5-1?

2. How do you think changing the direction of the current flow by reversing the connections to the battery in Figure 5-1 will change the rankings in (1)?

3. How can you compare the currents in the circuits in Figure 5-1 experimentally? List the equipment you will need.

4. Define *series* and *parallel* connections. Sketch two light bulbs connected to a battery in series and two light bulbs connected to a battery in parallel.

5. Predict how the brightness of bulb D will change when the switch is closed in Figure 5-6.

6. Predict how the current flowing through the battery will change when the switch is closed in Figure 5-6.

LAB 5:
CURRENT IN SIMPLE DC CIRCUITS*

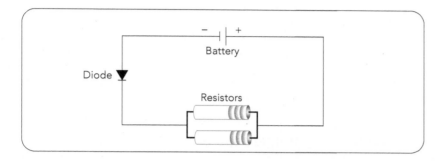

If it's green and it wiggles, it's biology.

If it stinks, it's chemistry.

If it doesn't work, it's physics.

If it's incomprehensible, it's mathematics.

If it doesn't make sense, it's either economics or psychology.

—From A. Bloch's
Murphy's Law Book 3

OBJECTIVES

• To understand how current flows in a circuit when a battery lights a bulb.

• To understand what a series connection is in an electric circuit.

• To understand the relationship between the currents in all parts of a series circuit.

• To understand what a parallel connection is in an electric circuit.

• To understand the relationship between the currents in all parts of a parallel circuit.

• To begin to understand the concept of resistance.

OVERVIEW

In Lab 4 you saw that when there is an electric current through a light bulb, the bulb lights. You also saw that to cause current to flow through a bulb, you must have a complete circuit that includes a voltage source such as a battery. You found that in a simple circuit containing a battery, a single bulb and wires connecting them, current will only flow when there is a complete path from the positive terminal of the battery, through the connecting wire to the bulb, through the bulb, through the connecting wire to the negative terminal of the battery.

By measuring the current at different points in a simple circuit consisting of a bulb, a battery, and connecting wires, you discovered a model for current flow,

---

*Some of the activities in this lab have been adapted from those designed by the Physics Education Group at the University of Washington, as adapted for Workshop Physics, Module 4.

namely, that the electric current is the same in all parts of the circuit. By measuring the current and voltage in this circuit and adding a second bulb, you also discovered that a fresh battery maintains essentially the same voltage whether it is connected to one light bulb or two.

You also observed that the current was smaller when a second bulb was added to the circuit. This led us to introduce the concept of *resistance* of a circuit element such as a bulb. The total resistance of the elements in a circuit determines how much current is provided by the battery.

In this lab you will examine more complicated circuits than a single bulb connected to a single battery. You will compare the currents through different parts of these circuits by comparing the brightness of the bulbs, and also by measuring these currents using current sensors. Later, in Lab 6, you will examine the role of the battery in causing a current in a circuit, and compare the potential differences (voltages) across different parts of your circuits.

INVESTIGATION 1: CURRENT IN SERIES CIRCUITS

In the next series of activities you will be asked to make a number of predictions about the current in several different circuits and then to compare them with actual observations. Whenever your experimental observations disagree with your predictions you should try to develop new concepts about how circuits with batteries and bulbs actually work. To make the required observations you will need the following items:

- computer-based laboratory system
- two current sensors
- *RealTime Physics Electricity and Magnetism* experiment configuration files
- a very fresh, alkaline 1.5-V D cell battery with holder
- 6 wires with alligator clip leads
- 4 #14 bulbs in sockets
- contact switch

Prediction 1-1: Consider the two circuits shown in Figure 5-1.

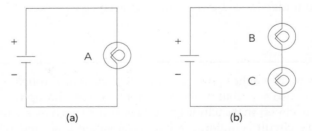

(a)   (b)

Figure 5-1: Two different circuits: (a) a battery with a single bulb, and (b) an identical battery with two bulbs B and C that are identical to bulb A.

Predict the relative brightness of the three bulbs, A, B and C shown in Figure 5-1 from brightest to dimmest. (Recall that in the last lab you observed that a fresh battery maintains essentially the same *voltage* across its terminals whether one or two bulbs are connected to it.)

If two or more bulbs are equal in brightness, indicate this in your response. Explain the reasons for your rankings.

> **Hint:** Helpful symbols are >, "is greater than"; <, "is less than"; =, "is equal to." For example, B > C > A.

Activity 1-1: The Relative Brightness of Bulbs

1. Now connect the circuits, observe the relative brightness of the bulbs, and rank the bulbs in the order of the brightness you actually observed.

> **Comment:** These activities assume *identical* bulbs. Differences in brightness may arise if the bulbs are not exactly identical. In this and later activities, to determine whether a difference in brightness is caused by a difference in the currents through the bulbs or by a difference in the bulbs, you should exchange the bulbs.
>
> Sometimes a bulb will not light noticeably, even if there is a small but significant current through it. If a bulb is really off, that is, if there is no current through it, then unscrewing the bulb will not affect the rest of the circuit. To verify whether a nonglowing bulb actually has a current through it, unscrew the bulb and see if anything else in the circuit changes.

Ranking of the bulbs:

Question 1-1: Did your observations agree with your predictions? If not, explain what assumptions you were making that now seem false.

Prediction 1-2: What do you predict will happen to the brightness of bulbs A, B, and C in Figure 5-1 if the battery is connected to the bulbs with its terminals reversed? Explain the reason(s) for your prediction.

2. Test your prediction. Reverse the terminals of the battery in each of your circuits.

Question 1-2: What do you observe? Does this agree with your prediction? Did you make any false assumptions? Explain.

Question 1-3: Can you tell the direction of the current through the circuit by just looking at the brightness of the bulbs without knowing how the battery is hooked up? Explain.

How does each bulb affect the current in a circuit? Does current get used up as it passes through a bulb? Is the current in the two-bulb circuit the same as, more than, or less than that in the single-bulb circuit? First make predictions and then observe what happens.

Prediction 1-3: What would you predict about the relative amount of current going through each bulb in Figures 5-1a and b? Write down your predicted rankings of the currents through bulbs A, B, and C, and explain your reasoning.

Activity 1-2: Current in a Simple Circuit with Bulbs

You can test your prediction by using current sensors. Recall from Lab 1 that to measure the current through a bulb, a current sensor must be connected so that the current through the current sensor is the same as the current through the bulb. Convince yourself that the current sensors shown in Figure 5-2 measure the currents described in the figure caption.

> **Comment:** In carrying out your measurements, it is important to realize that the measurements made by the current sensors are only as good as their calibrations. Small differences in calibration can result in small differences in readings.

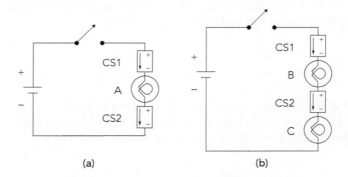

(a)   (b)

Figure 5-2: Current sensors connected to measure the current through bulbs. In circuit (a), CS1 measures the current flowing into bulb A, and CS2 measures the current flowing out of bulb A. In circuit (b), CS1 measures the current flowing into bulb B while CS2 measures the current flowing out of bulb B and into bulb C.

1. Open the experiment **Two Currents (L05A1-2)** to display the two sets of current vs. time axes that follow.

2. **Calibrate** the current sensors, or load the calibration. **Zero** the sensors with them disconnected from the circuit.

3. Connect circuit (a) in Figure 5-2.

4. **Begin graphing,** close the switch for a second or so, open it for a second or so, and then close it again. Sketch your graphs on the axes that follow or **print** your graphs and affix them over the axes.

5. Use the **analysis feature** of the software to measure the currents flowing into and out of bulb A when the switch is closed:

Current into bulb A:———   Current out of bulb A:———

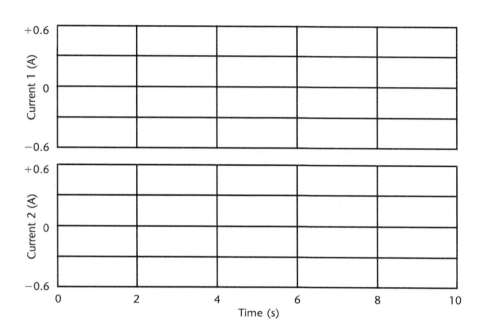

Question 1-4: Are the currents flowing into and out of bulb A equal or is one significantly larger (i.e., do they differ by more than a few percent)? What can you say about the directions of the currents? Is this what you expected? Explain.

6. Connect circuit (b) in Figure 5-2. **Begin graphing** current as above, and record the measured values of the currents.

Current through bulb B:——— Current through bulb C:———

7. Sketch the graphs on the axes that follow, or **print** and affix over the axes.

Question 1-5: Consider your observation of the circuit with bulbs B and C in it. Is current "used up" in the first bulb, or is it the same in both bulbs? Explain based on your observations.

Question 1-6: Is the ranking of the currents in bulbs A, B, and C what you predicted? If not, can you explain what assumptions you were making that now seem false?

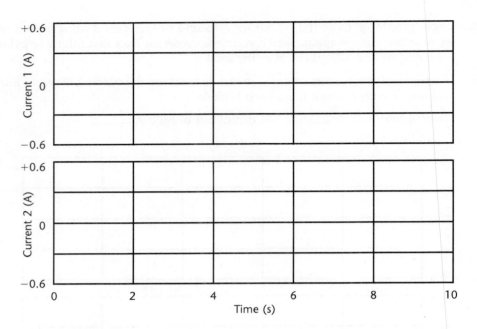

Question 1-7: Based on your observations, how is the brightness of a bulb related to the current flowing through it?

Question 1-8: How does the current produced by the battery in the single bulb circuit (Figure 5-1a) compare to that produced by the battery with two bulbs that are connected as in Figure 5-1b? Does the addition of this second bulb in this manner affect the current through the original bulb? Explain.

Question 1-9: Suppose you think of the bulb as providing a *resistance* to the current in a circuit, rather than something that uses up current. How do you expect the total resistance in a circuit to be affected by the addition of more bulbs in the manner shown in Figure 5-1b?

Question 1-10: Formulate a rule for predicting whether current increases or decreases as the total resistance of the circuit is increased.

> **Comment:** The rule you have formulated based on your observations with bulbs may be *qualitatively* correct—correctly predicting an increase or decrease in current—but it won't be *quantitatively* correct. That is, it won't allow you to predict the exact sizes of the currents correctly. This is because the resistance of a bulb to current changes as the current flowing through the bulb changes. You will explore this in more detail in Lab 6.
>
> Another common circuit element is a *resistor*. A resistor has a constant resistance to current regardless of how large the current flowing through it. In the next activity you will reformulate your rule using resistors.

First a prediction.

Prediction 1-4: Consider the circuit diagrams in Figure 5-3 in which the light bulbs in Figure 5-1 have been replaced by identical resistors.

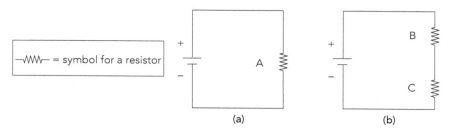

Figure 5-3: Two different circuits: (a) a battery with a single resistor, and (b) a battery with two resistors identical to the one in (a).

What would you predict about the relative amount of current flowing through each resistor in Figures 5-3a and b? Write down your predicted rankings of the currents flowing through resistors A, B, and C, and explain your reasoning. (Remember that a resistor has a constant resistance to current regardless of the current flowing through it.)

In addition to the equipment listed above, you will need the following to test your predictions:

* two 10-ohm resistors

Activity 1-3: Current in a Simple Circuit with Resistors

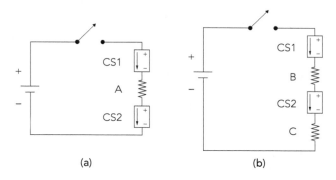

Figure 5-4: Current sensors connected to measure the currents through resistors. In circuit (a), CS1 measures the current into resistor A, and CS2 measures the current out of resistor A. In circuit (b), CS1 measures the current into resistor B, while CS2 measures the current out of resistor B and into resistor C.

1. Continue to use the experiment file **Two Currents (L05A1-2).**

2. **Calibrate** the current sensors, or load the calibration if this has not already been done. **Zero** the sensors with them disconnected from the circuit.

3. Connect circuit (a) in Figure 5-4.

4. Use the current sensors and the **analysis feature** in the software to measure the current flowing through resistor A in circuit 5-4a and the currents flowing through resistors B and C in circuit 5-4b.

   Current flowing through resistor A:_____

   Current flowing through resistor B:_____

   Current flowing through resistor C:_____

Question 1-11: Is the ranking of the currents through resistors A, B, and C what you predicted? If not, can you explain what assumptions you were making that now seem false?

Question 1-12: How does the current produced by the battery in the single resistor circuit (Figure 5-3a) compare to that produced by the battery with two resistors connected as in Figure 5-3b? Does the addition of this second resistor in this manner affect the current flowing through the original resistor? Explain.

Question 1-13: How did your observations with resistors differ from your observations with bulbs in Activity 1-2?

Question 1-14: Reformulate a more quantitative rule for predicting *how* the current supplied by the battery decreases as more *resistors* are connected in the circuit as in Figure 5-3b.

Question 1-15: Is your rule in Question 1-14 also correct for bulbs connected as in Figures 5-1a and b? Explain.

INVESTIGATION 2: CURRENT IN PARALLEL CIRCUITS

There are two basic ways to connect resistors, bulbs, or other elements in a circuit: *series* and *parallel*. So far you have been connecting bulbs and resistors *in series*. To make predictions involving more complicated circuits, we need to have a more precise definition of series and parallel. These are summarized below.

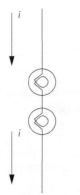

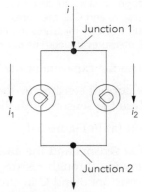

Series connection:
Two resistors are in series if they are connected so that the same current that passes through one passes through the other.

Parallel connection:
Two resistors are in parallel if their terminals are connected so that at each junction one terminal of one resistor is directly connected to one terminal of the other.

It is important to keep in mind that in more complex circuits, say with three or more elements, not every element is necessarily connected in series or in parallel with other elements.

Let's compare the behavior of a circuit with two bulbs wired in parallel to the circuit with a single bulb (see Figure 5-5).

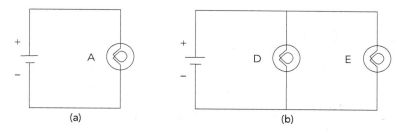

Figure 5-5: Two different circuits: (a) a single-bulb circuit and (b) a circuit with two bulbs identical to the one in (a) connected *in parallel* to each other and *in parallel* to the battery.

Note that if bulbs A, D, and E are identical, then the circuit in Figure 5-6 is equivalent to circuit 5-5a when the switch is open (as shown) and equivalent to circuit 5-5b when the switch is closed.

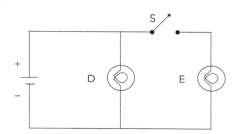

Figure 5-6: When the switch is open, only bulb D is connected to the battery. When the switch is closed, bulbs D and E are connected *in parallel* to each other and *in parallel* to the battery.

Question 2-1: Explain how you know that the caption of Figure 5-6 correctly describes the circuit.

Prediction 2-1: Predict how the brightness of bulbs D and E in the parallel circuit of Figure 5-5b will compare to bulb A in the single bulb circuit of Figure 5-5a. How will D and E compare with each other? Rank the brightness of all three bulbs. Explain the reasons for your predictions.

To test this and other predictions, you will need:

- computer-based laboratory system
- two current sensors
- *RealTime Physics Electricity and Magnetism* experiment configuration files

- a very fresh, alkaline 1.5-V D cell battery with holder
- 8 wires with alligator clip leads
- 3 #14 bulbs in sockets
- contact switch

Activity 2-1: Brightness of Bulbs in a Parallel Circuit

Set up the circuit in Figure 5-6, and describe your observed rankings for the brightness of bulb D with the switch open, and D and E with the switch closed.

Question 2-2: Did the observed rankings agree with your prediction? If not, can you explain what assumptions you made that now seem false?

Prediction 2-2: Based on your observations of brightness, what do you predict about the relative amount of current flowing through each bulb in a parallel connection, i.e., bulbs D and E in Figure 5-5b?

Prediction 2-3: Based on your observations of brightness, how do you think that closing the switch in Figure 5-6 affects the current flowing through bulb D?

Activity 2-2: Current in Parallel Branches

You can test Predictions 2-2 and 2-3 by connecting current probes to measure the currents through bulbs D and E.

1. Open the experiment file called **Two Currents (L05A1-2),** if it is not already opened.
2. **Calibrate** the current sensors, or load the calibration, if this hasn't already been done. **Zero** the sensors with them disconnected from the circuit.
3. Connect the circuit shown in Figure 5-7.

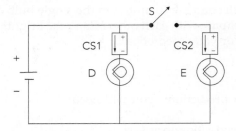

Figure 5-7: Current sensors connected to measure the current flowing through bulb D and the current flowing through bulb E.

4. **Begin graphing** the currents flowing through both sensors, then close the switch for a second or so, open it for a second or so, and then close it again.

5. Sketch the graphs on the axes below, or **print** them and affix them over the axes.

6. Use the **analysis feature** of the software to measure both currents.

   Switch open: Current flowing through bulb D:_____
   Current flowing through bulb E:_____

   Switch closed: Current flowing through bulb D:_____
   Current flowing through bulb E:_____

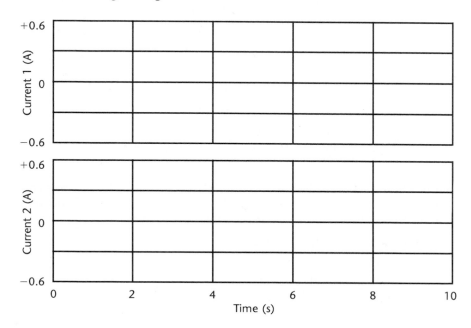

Question 2-3: Based on your graphs and measurements, were the currents flowing through bulbs D and E what you predicted based on their brightness? If not, can you now explain why your prediction was incorrect?

Question 2-4: Did closing the switch and connecting bulb E *in parallel* with bulb D significantly affect the current flowing through bulb D? How do you know? [**Note:** You are making a *very significant* change in the circuit. Think about whether the new current flowing through D when the switch is closed reflects this.]

You have already seen in Lab 4 that the voltage maintained by a battery doesn't change appreciably no matter what is connected to it. But what about the current flowing through the battery? Is it always the same no matter what is connected to it, or does it change depending on the circuit? (Is the current flowing through the battery the same whether the switch in Figure 5-6 is open or closed?) This is what you will investigate next.

Prediction 2-4: Based on your observations of the brightness of bulbs D and E in Activity 2-2, what do you predict about the amount of current flowing through the battery in the parallel bulb circuit (Figure 5-5b) compared to that flowing through the single bulb circuit (Figure 5-5a)? Explain.

Activity 2-3: Current Flowing Through the Battery

1. Test your prediction with the circuit shown in Figure 5-8. Use the same experiment file, **Two Currents (L05A1-2),** as in the previous activities.

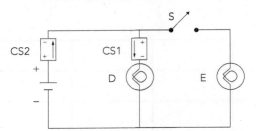

Figure 5-8: Current sensors connected to measure the current through the battery and the current through bulb D.

2. **Begin graphing** while closing and opening the switch as before. Sketch your graphs on the axes that follow, or **print** and affix over the axes. Label on your graphs when the switch is open and when it is closed.

3. Measure the currents flowing through the battery and through bulb D.

   Switch open: Current flowing through battery:____
   Current flowing through bulb D:____

   Switch closed: Current flowing through battery:____
   Current flowing through bulb D:____

Question 2-5: Describe how the connection of current sensors in Figure 5-8 differs from that in Figure 5-7. How do you know that sensor 2 is measuring the current flowing through the battery?

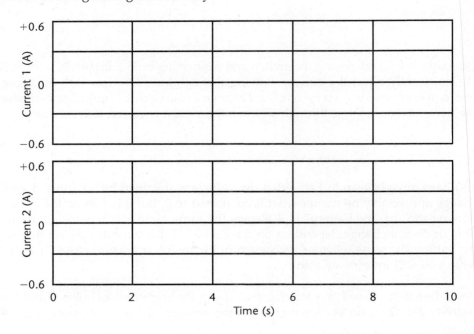

Question 2-6: Use your observations to formulate a rule to predict how the current flowing through a battery will change as the number of bulbs connected *in parallel* increases. Can you explain why?

Question 2-7: Compare your rule in Question 2-6 to the rule you stated in Questions 1-10 and 1-14 relating the current flowing through the battery to the total *resistance* of the circuit. Does adding more bulbs in parallel increase, decrease, or not change the total *resistance* of the circuit? Explain.

Question 2-8: Can you explain your answer to Question 2-7 in terms of the number of paths for current available in the circuit? Explain.

Question 2-9: Considering your experiences with series and parallel circuits in Investigations 1 and 2, does the current flowing through the battery depend only on the number of bulbs or resistors in the circuit, or does the arrangement of the circuit elements matter?

Question 2-10: Since current and resistance are related, does the resistance depend just on the number of bulbs or resistors, or does it depend on the arrangement of the circuit elements as well? Explain.

INVESTIGATION 3: MORE COMPLEX SERIES AND PARALLEL CIRCUITS

Now you can apply your knowledge to some more complex circuits. Consider the circuit consisting of a battery and two bulbs, A and B, in series shown in Figure 5-9a. What will happen if you add a third bulb, C, in parallel with bulb B as shown in Figure 5-9b? You should be able to predict the relative brightness of A, B, and C based on previous observations. The tough question is: how does the brightness of A change when C is connected in parallel to B?

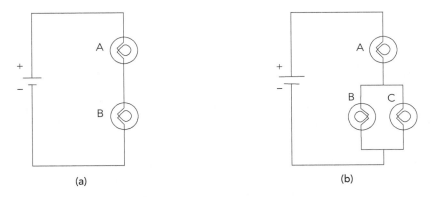

(a)   (b)

Figure 5-9: Two circuits with identical batteries and bulbs A, B, and C.

Question 3-1: In Figure 5-9b is bulb A in series with bulb B? with bulb C? or with a combination of bulbs B and C? (You may want to go back to the definitions of series and parallel connections at the beginning of Investigation 2.)

Question 3-2: In Figure 5-9b are bulbs B and C connected in series or in parallel with each other, or neither? Explain.

Question 3-3: Is the resistance of the combination of bulbs B and C larger than, smaller than, or the same as bulb B alone? Explain.

Question 3-4: Is the resistance of the combination A, B, and C in Figure 5-9b larger than, smaller than, or the same as the combination of A and B in Figure 5-9a? Explain.

Prediction 3-1: Predict how the current flowing through bulb A will change, if at all, when circuit 5-9a is changed to 5-9b (when bulb C is added in parallel to bulb B). What will happen to the brightness of bulb A? Explain the reasons for your predictions.

Prediction 3-2: Predict how the current flowing through bulb B will change, if at all, when circuit 5-9a is changed to 5-9b (when bulb C is added in parallel to bulb B). What will happen to the brightness of bulb B? Explain the reasons for your predictions.

Prediction 3-3: Also predict the relative rankings of brightness for all the bulbs, A, B, and C, after bulb C is in the circuit. Explain the reasons for your predictions.

Activity 3-1: A More Complex Circuit

1. Set up the circuit shown in Figure 5-10a. Convince yourself that this circuit is identical to Figure 5-9a when the switch, S, is open, and to Figure 5-9b when the switch is closed.

2. Observe the brightness of bulbs A and B when the switch is open, and then the brightness of the three bulbs when the switch is closed. Compare the brightness of bulb A with the switch open and closed, and rank the brightness of bulbs A, B, and C with the switch closed.

Question 3-5: If you did not observe what you predicted about the brightness, what changes do you need to make in your reasoning? Explain.

3. Connect the two current sensors as shown in Figure 5-10b. Open the experiment file called **Two Currents (L05A1-2),** if it is not already opened.

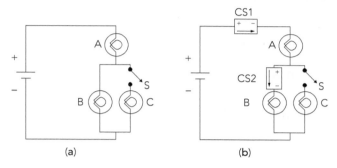

(a)　　　　　　　　(b)

Figure 5-10: (a) Circuit equivalent to Figure 5-9a when the switch, S, is open and to Figure 5-9b when the switch is closed. (b) Same circuit with current sensors connected to measure the current through bulb A (CS1) and the current through bulb B (CS2).

4. **Begin graphing** and observe what happens to the current flowing through bulb A (through the battery) and the current flowing through bulb B when the switch is opened and closed.

Question 3-6: What happens to the current flowing through the battery and through bulbs A and B when bulb C is added in parallel with bulb B? What do you conclude happens to the total resistance in the circuit? Explain.

If you have additional time, do some or all of the following Extensions to examine some more complex circuits

Extension 3-2: An Even More Complex Circuit

Let's look at a somewhat more complicated circuit to see how series and parallel parts of a complex circuit affect one another. The circuit is shown in Figure 5-11.

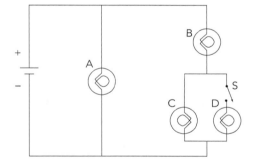

Figure 5-11: A complex circuit with series and parallel connections.

Question E3-7:   When switch S in Fig. 5-11 is open, which bulbs are connected in parallel with each other? (If you need to, review the definitions of series and parallel at the beginning of Investigation 2 before answering.)

Is A parallel to B?

Is A parallel to C?

Is C parallel to D?

Is A parallel to the combination of B and C?

Question E3-8:   When switch S is open, which bulbs are connected in series with each other?

Is A in series with B?

Is A in series with C?

Is B in series with C?

Question E3-9:   When switch S is closed, which bulb(s) are connected in parallel with A?

Question E3-10:   When switch S is closed, which bulb(s) are connected in series with B?

Prediction E3-4:   Predict the effect on the current flowing through bulb A for each of the following separate alterations in the circuit:

a.   unscrewing bulb B

b.   closing switch S

Prediction E3-5:   Predict the effect on the current flowing through bulb B of each of the following separate alterations in the circuit:

a.   unscrewing bulb A

b.   adding another bulb in series with bulb A

Connect the circuit in Figure 5-11, and observe the effect of each of the alterations in Predictions E3-4 and E3-5 on the brightness of each bulb. Describe your observations.

Question E3-11: Compare your results with your predictions. How do you account for any differences between your predictions and observations?

Question E3-12: In this circuit, two parallel branches are connected *directly across* a battery. For this type of connection, what do you conclude about the effect of changes in one parallel branch on the current flowing in the other?

Extension 3-3: Series and Parallel Networks

Now let's practice with some more complicated series and parallel circuits. Suppose you had three boxes, labeled A, B, and C, each having two terminals. The arrangements of resistors in the boxes are shown in Figure 5-12.

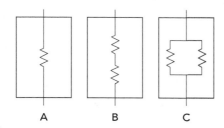

Figure 5-12: Parallel and series circuits.

Consider the five circuits shown in Figure 5-13 in answering the following questions.

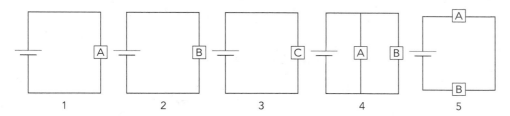

Figure 5-13: Circuits featuring parallel and series connections.

Question E3-13: For each of the circuits in Figure 5-13, sketch a standard circuit diagram showing all the resistors in the circuit. In each diagram, label each of the resistors in the circuit with a number. Then describe which resistors or combination of resistors are in series and parallel with each other.

Question E3-14: Rank the five circuits in Figure 5-13 by their total resistance. Which has the most resistance? The least resistance? Explain your reasoning.

Question E3-15: Rank each of the circuits in Figure 5-13 according to the total current flowing through the battery. Explain your reasoning.

You can now test your understanding of current and resistance on another puzzling circuit.

Extension 3-4: The Puzzle Problem

Question E3-16: Use reasoning based on your model of electric current to predict the relative brightness of each of the bulbs shown in Figure 5-14. Explain the reasons for your rankings.

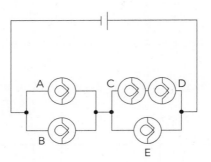

Figure 5-14: A complicated battery and bulb circuit.

Name_____ Date_____ Partners_____

HOMEWORK FOR LAB 5
CURRENT IN SIMPLE DIRECT CURRENT CIRCUITS

1. Which of the three circuits shown below, if any, are the same electrically? Which are different? Explain your answers.

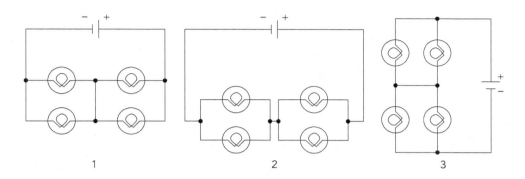

2. Consider the two messy circuit diagrams 1 and 2 below.

   a. Identify which of the nice neat circuit diagrams below (A, B, C, or D) corresponds to circuit 1. Explain the reasons for your answer.

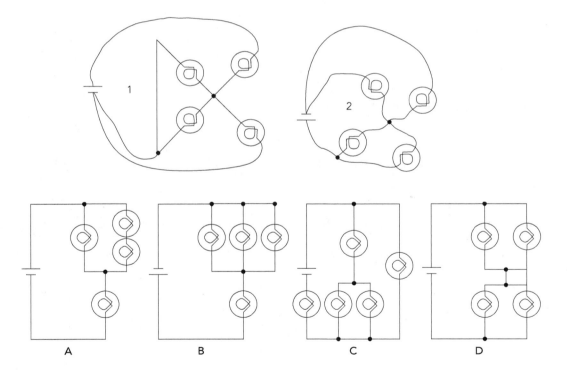

   b. Which circuit diagram (A, B, C, or D) corresponds to circuit 2? Explain the reasons for your answer.

3. Three of the circuits drawn below are electrically equivalent and one is not.

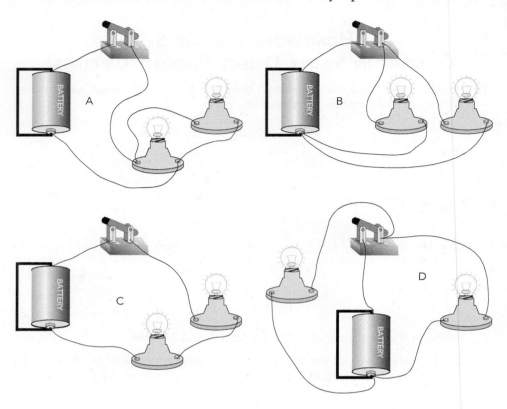

a. Which circuit is not like the others? Explain how it is different.

b. Which circuits represent parallel arrangements for the bulbs? Which represent series arrangements?

c. In the boxes below, draw neat circuit diagrams for each of the arrangements.

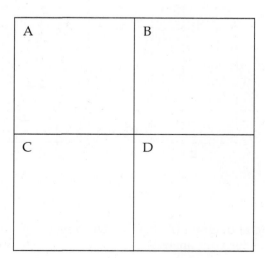

4. Use the model for electric current to rank the resistor networks shown below in order by resistance from largest to smallest. Explain your reasoning.

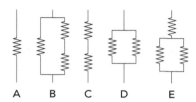

A   B   C   D   E

5. If a battery were connected to each of the circuits in Question 4, in which network would the current flowing through the battery be the largest? The smallest? Explain your reasoning.

6. The diagram below shows a typical household circuit. The appliances (lights, television, toaster, etc.) are represented by boxes labeled 1, 2, 3, and so on. The fuse, or circuit breaker, shown in the diagram is a switch intended to shut off the circuit automatically if the wires become too hot because the current flowing in the circuit is too large.

> **Note:** Although houses in the United States use alternating current (AC), which differs in some important ways from the direct current (DC) we have been studying, you can still use the model you developed in answering these questions.

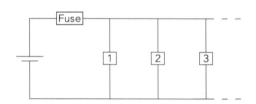

   a. What happens to the current flowing through the fuse when more appliances are added to the circuit? Describe evidence from this lab for your answer.

   b. Does the current flowing through element 1 change when elements 2 and 3 are added to the circuit? Describe evidence from this lab for your answer.

   c. Is this model consistent with your observations of everyday household electricity? For example, what happens to the brightness of a light bulb in a room when a second one is turned on?

d. What may happen to the fuse if too many appliances are added to the circuit? Why?

e. What kind of circuit connection for elements 1, 2, and 3 is shown in the diagram?

7. Consider the circuit shown on the right.

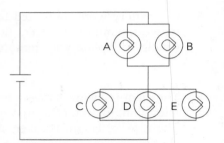

   a. Are the bulbs C, D, and E connected in series, parallel, or neither? Explain.

   b. Rank the bulbs in order of brightness. Use the symbols =, <, and >. Explain your ranking.

   c. How will the brightness of bulbs A and B change if bulb C is unscrewed? Will the result be different if bulb D or E is unscrewed instead? Explain.

8. Consider the circuit shown on the right. Rank the brightness of the bulbs in the circuit. Use the symbols =, <, and >. Explain your ranking.

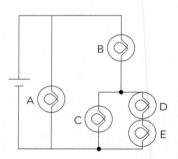

9. In the two circuits below, the batteries and all bulbs are identical. Compare the current flowing in the circuit on the left to the current flowing in the circuit on the right. Be as quantitative as possible.

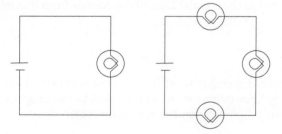

10. In the two circuits below, the batteries and all resistors are identical. Compare the current flowing in the circuit on the left to the current flowing in the circuit on the right. Be as quantitative as possible.

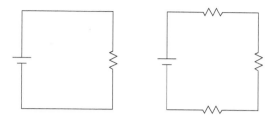

Is your answer the same as in Question 9? Explain any differences.

Name_____   Date_____

PRE-LAB PREPARATION SHEET FOR
LAB 6—VOLTAGE IN SIMPLE DC CIRCUITS AND OHM'S LAW

(Due at beginning of lab)

Directions:

Read over Lab 6 and then answer the following questions about the procedures.

1. What do you predict for the brightness of the bulbs in Figure 6-1a and Figure 6-1c?

2. What do you predict for the brightness of the bulbs in Figure 6-2 and Figure 6-1a?

3. How will you measure the voltages between points 1 and 2 in the three circuits shown in Figure 6-3? What equipment will you use?

4. What do you predict will happen to the voltage across the battery in Figure 6-7a when you close the switch?

5. What is the function of the power supply in Figures 6-12 and 6-13?

6. How will you determine the quantitative relationship between the voltage across a resistor and the current flowing through it? What equipment will you use?

LAB 6:
VOLTAGE IN SIMPLE DC CIRCUITS AND OHM'S LAW*

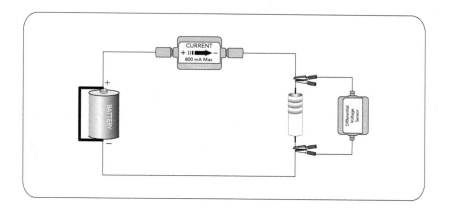

*I have a strong resistance to understanding
the relationship between voltage and current.*
—Anonymous introductory physics student

OBJECTIVES

- To learn to apply the concept of potential difference (voltage) to explain the action of a battery in a circuit.

- To understand how potential difference (voltage) is distributed in different parts of a series circuit.

- To understand how potential difference (voltage) is distributed in different parts of a parallel circuit.

- To understand the quantitative relationship between potential difference and current for a resistor (Ohm's law).

OVERVIEW

In the last two labs you explored currents flowing at different points in series and parallel circuits. You saw that in a series circuit, *the same current flows through all elements.* You also saw that in a parallel circuit, *the current divides among the branches so that the total current flowing through the battery equals the sum of the currents flowing through each of the branches.*

You have also observed that when two or more parallel branches are connected directly across a battery, making a change in one branch does not affect the current flowing in the other branch(es), while changing one part of a series circuit changes the current flowing in all parts of that series circuit.

---

*Some of the activities in this lab have been adapted from those designed by the Physics Education Group at the University of Washington, as adapted for Workship Physics Module 4.

In carrying out these observations of series and parallel circuits, you have seen that connecting light bulbs in series results in a larger resistance to current flow and therefore a smaller current, while a parallel connection results in a smaller resistance and larger current.

In this lab, you will first examine the role of the battery in causing a current to flow in a circuit. You will then compare the potential differences (voltages) across different parts of series and parallel circuits.

Based on your previous observations, you probably associate a larger resistance connected to a battery with a smaller current, and a smaller resistance with a larger current. In the last part of this lab you will explore the quantitative relationship between the current flowing through a *resistor* and the potential difference (voltage) across the resistor. This relationship is known as Ohm's law.

INVESTIGATION 1: BATTERIES AND VOLTAGES IN SERIES CIRCUITS

So far you have developed a current model and the concept of resistance to explain the relative brightness of bulbs in simple circuits. Your model says that when a battery is connected to a complete circuit, there is a current flow. For a given battery, the magnitude of the current depends on the total resistance of the circuit. In this investigation you will explore batteries and the potential differences (voltages) between various points in circuits.

To do this you will need the following items:

- computer-based laboratory system
- two voltage sensors
- *RealTime Physics Electricity and Magnetism* experiment configuration files
- 2 very fresh, alkaline 1.5-V D cell batteries and holders
- 6 wires with alligator clip leads
- 4 #14 bulbs in sockets
- contact switch

You have already seen what happens to the brightness of the bulb in circuit 6-1a if you add a second bulb in series as shown in circuit 6-1b. The two bulbs are not as bright as the original bulb. We concluded that the resistance of the circuit is larger, resulting in less current flowing through the bulbs.

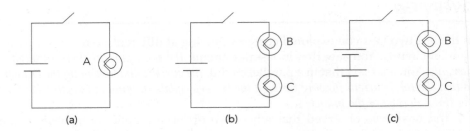

Figure 6-1: Series circuits with (a) one battery and one bulb, (b) one battery and two bulbs, and (c) two batteries and two bulbs. (All batteries and all bulbs are identical.)

Prediction 1-1: What do you predict would happen to the brightness of the bulbs if you connected a second battery in series with the first at the same time you added the second bulb as in Figure 6-1c? How would the brightness of bulb A in circuit 6-1a compare to bulb B in circuit 6-1c? To bulb C?

Activity 1-1: Battery Action

1. Connect the circuit in Figure 6-1a, and observe the brightness of the bulb.

2. Now connect the circuit in Figure 6-1c. (Be sure that the batteries are connected *in series*—the positive terminal of one must be connected to the negative terminal of the other.)

Question 1-1: Compare the brightness of each of the bulbs to the single-bulb circuit.

Question 1-2: What do you conclude about the current flowing in the two-bulb, two-battery circuit as compared to the single-bulb, single-battery circuit? Explain.

Question 1-3: What happens to the resistance of a circuit as more bulbs are added in series? What must you do to keep the current from decreasing?

Prediction 1-2: What do you predict about the brightness of bulb D in Figure 6-2 compared to bulb A in Figure 6-1a? Explain your prediction.

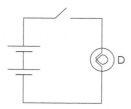

Figure 6-2: Series circuit with two batteries and one bulb.

3. Connect the circuit in Figure 6-2. *Close the switch for only a moment to observe the brightness of the bulb—otherwise, you will burn out the bulb.*

Question 1-4: Compare the brightness of bulb D to the single-bulb circuit with only one battery (bulb A in Figure 6-1a).

Question 1-5: How does increasing the number of batteries connected in series affect the current flowing in a series circuit?

When a battery is fresh, the voltage marked on it is actually a measure of the *emf* (*electromotive force*) or electric *potential difference* between its terminals. *Voltage* is an informal term for emf or potential difference. If you want to talk to physicists, you should refer to potential difference. Communicating with a salesperson at the local Radio Shack store is another story. There you would probably refer to voltage. We will use the two terms interchangeably.

Let's explore the potential differences of batteries and bulbs in series and parallel circuits to see if we can come up with rules for them as we did earlier for currents.

How do the potential differences of batteries add when the batteries are connected in series or parallel? Figure 6-3 shows a single battery, two batteries identical to it connected in series, and two batteries identical to it connected in parallel.

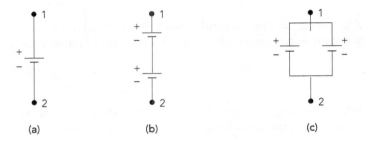

(a)   (b)   (c)

Figure 6-3: Identical batteries: (a) single, (b) two connected in series, and (c) two connected in parallel.

Prediction 1-3: If the potential difference between points 1 and 2 in Figure 6-3a is known, predict the potential difference between points 1 and 2 in Figure 6-3b (series connection) and in Figure 6-3c (parallel connection). Explain your reasoning.

Activity 1-2: Batteries in Series and Parallel

You can measure potential differences with voltage sensors connected as shown in Figure 6-4.

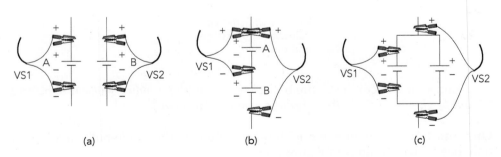

(a)   (b)   (c)

Figure 6-4: Voltage sensors connected to measure the potential difference across (a) two single batteries, (b) a single battery and two batteries connected in series, and (c) a single battery and two batteries connected in parallel.

REALTIME PHYSICS: ELECTRICITY AND MAGNETISM

1. Open the experiment file called **Batteries (L06A1-2)**.

2. **Calibrate** the sensors or **load the calibration**. **Zero** the sensors with the red and black alligator clips connected together.

3. Connect voltage sensor 1 across a single battery (as in Figure 6-4a), and voltage sensor 2 across the other identical battery.

4. Record the voltage measured for each battery below.

   Voltage of battery A:_____   Voltage of battery B:_____

Question 1-6: How do your measured values agree with those marked on the batteries?

5. Now connect the batteries in series as in Figure 6-4b, and connect sensor 1 to measure the potential difference across battery A and sensor 2 to measure the potential difference across the series combination of the two batteries. Record your measured values below.

   Voltage of battery A:_____   Voltage of A and B in series:_____

Question 1-7: Do your measured values agree with your predictions? Explain any differences?

6. Now connect the batteries in parallel as in Figure 6-4c, and connect sensor 1 to measure the potential difference across battery A and sensor 2 to measure the potential difference across the parallel combination of the two batteries. Record your measured values below.

   Voltage of battery A:_____   Voltage of A and B in parallel:_____

Question 1-8: Do your measured values agree with your predictions? Explain any differences.

Question 1-9: Make up a rule for finding the combined voltage of a number of batteries connected in series.

Question 1-10: Make up a rule for finding the combined voltage of a number of identical batteries connected in parallel.

You can now explore the potential difference across different parts of a simple series circuit. Let's begin with the circuit with two bulbs in series with a battery, which you looked at before in Lab 5, Activities 1-1 and 1-2. It is shown in Figure 6-5a.

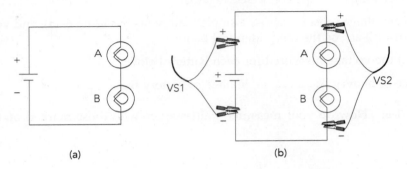

Figure 6-5: (a) A series circuit with one battery and two bulbs, and (b) the same circuit with voltage sensor 1 connected to measure the potential difference across the battery and sensor 2 connected to measure the potential difference across the series combination of bulbs A and B.

Prediction 1-4: If bulbs A and B are identical, predict how the potential difference (voltage) across bulb A in Figure 6-5b will compare to the potential difference across the battery. How about bulb B? How about the potential difference across the series combination of bulbs A and B—how will this compare to the voltage across the battery?

Test your prediction.

Activity 1-3: Voltages in Series Circuits

1. Open the experiment file called **Batteries (L06A1-2)**, if it is not already open.

2. **Calibrate** the voltage sensors or **load the calibration**, if this has not already been done. **Zero** both sensors with the red and black alligator clips connected together.

3. Connect the circuit shown in Figure 6-5b.

> **Comment:** In carrying out your measurements, remember that the measurements made by the voltage sensors are only as good as their calibration. Small differences in calibration can result in small differences in readings.

4. Measure the voltages, and record your readings below.

   Potential difference across the battery:_____

   Potential difference across bulbs A and B in series:_____

Question 1-11: How do the two potential differences compare? Did your observations agree with your predictions?

5. Connect the voltage sensors as in Figure 6-6 to measure the potential difference across bulb A and across bulb B. Record your measurements on the next page.

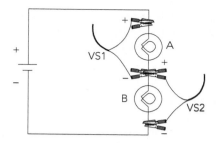

Figure 6-6: Connection of voltage sensors to measure the potential difference across bulb A and across bulb B.

Potential difference across bulb A:_____

Potential difference across bulb B:_____

Question 1-12: Did your measurements agree with your predictions?

Question 1-13: Formulate a rule for how potential differences across individual bulbs in a series connection combine to give the total potential difference across the series combination of the bulbs. How is this related to the potential difference of the battery?

INVESTIGATION 2: VOLTAGES IN PARALLEL CIRCUITS

You can also explore the potential differences across different parts of a simple *parallel* circuit. Let's begin with the circuit with two bulbs in parallel with a battery, which you looked at in Lab 5. It is shown in Figure 6-7a.

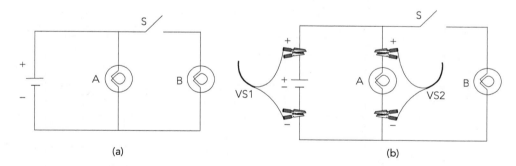

(a)   (b)

Figure 6-7: (a) Parallel circuit with two bulbs and a battery, and (b) the same circuit with voltage sensor 1 connected to measure the potential difference across the battery and sensor 2 connected to measure the potential difference across bulb A.

Prediction 2-1: What do you predict will happen to the potential difference across the battery when you close the switch in Figure 6-7a? Will it increase, decrease, or remain essentially the same? Explain.

Prediction 2-2: With the switch in Figure 6-7a closed, how will the potential difference across bulb A compare to the voltage of the battery? How will the potential difference across bulb B compare to the voltage of the battery?

To test your predictions you will need:

- computer-based laboratory system
- 2 voltage sensors
- *RealTime Physics Electricity and Magnetism* experiment configuration files
- very fresh, alkaline 1.5-V D cell battery with holder
- 6 alligator clip leads
- 2 #14 bulbs in sockets
- contact switch

Activity 2-1: Voltages in a Parallel Circuit

1. The experiment file **Batteries (L06A1-2)** should still be open, and the axes that follow should be on the screen.

2. **Calibrate** the voltage sensors or **load the calibration**, if this has not already been done. **Zero** both sensors with the red and black alligator clips connected together.

3. Connect the circuit shown in Figure 6-7b.

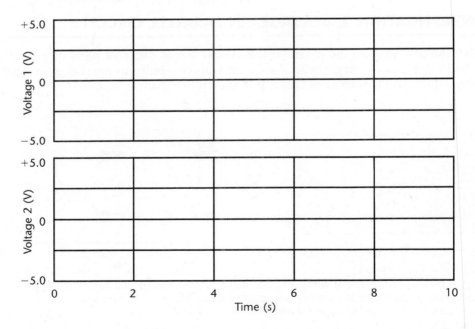

4. **Begin graphing**, and then close and open the switch as you've done before.

5. Sketch the graphs on the axes above, or **print** and affix them over the axes.

6. Read the voltages using the **analysis feature** of the software.

   Switch open: Voltage across battery:_____ Voltage across bulb A:_____

   Switch closed: Voltage across battery:_____ Voltage across bulb A:_____

Question 2-1: Did your measurements agree with your predictions? Did closing and opening the switch significantly affect the voltage across the battery (by more than several percent)? The voltage across bulb A?

7. Now connect the voltage sensors as shown in Figure 6-8, and graph and measure the voltages across bulbs A and B. Again close and open the switch while graphing.

8. Sketch your graphs or **print** them and affix them over the axes.

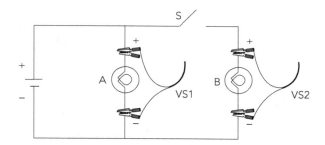

Figure 6-8: Voltage sensors connected to measure the potential differences across bulbs A and B.

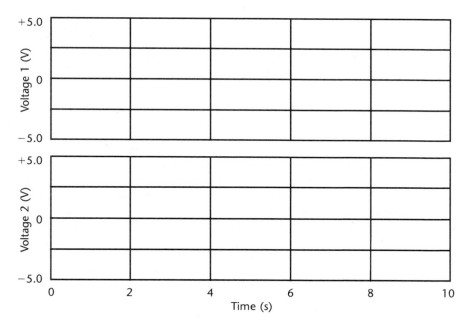

9. Record your measurements using the **analysis feature** of the software.

Switch open: Voltage across bulb A:_____ Voltage across bulb B:_____

Switch closed: Voltage across bulb A:_____ Voltage across bulb B:_____

Question 2-2: Did your measurements agree with your predictions? Did closing and opening the switch significantly affect the voltage across bulb A (by more than several percent)?

Question 2-3:   Did closing and opening the switch significantly affect the voltage across bulb B (by more than several percent)? Under what circumstances is there a potential difference across a bulb?

Question 2-4:   Based on your observations, formulate a rule for the potential differences across the different branches of a parallel circuit. How are these related to the voltage across the battery?

Question 2-5:   Is a fresh battery a constant current source (delivering essentially a fixed amount of current regardless of the circuit connected to it) or a constant voltage source (applying essentially a fixed potential difference regardless of the circuit connected to it), or neither? Explain based on your observations in this and the previous lab.

Question 2-6:   What is the voltage between two points on a short length of wire when there is no bulb, battery, or resistor between the points?

You have now observed several times in these activities that the voltage across a very fresh alkaline battery doesn't change significantly no matter what is connected to the battery (no matter how much current flows in the circuit). But, how does a battery that's not fresh behave?
  To answer this question, you'll need the materials you've been using along with

- 1.5-V D cell battery that is not fresh

- additional #14 bulbs and socket

- additional contact switch

Activity 2-2: Internal Resistance of a Battery

1. Open the experiment file **Internal Resistance (L06A2-2)** to measure voltage with voltage sensor 1 and current with current sensor 2.

2. **Calibrate** the sensors or **load the calibration**, if this has not already been done. **Zero** both sensors with the red and black alligator clips of the voltage sensor connected together, and nothing connected to the current sensor.

3. Connect the circuit shown in Figure 6-9.

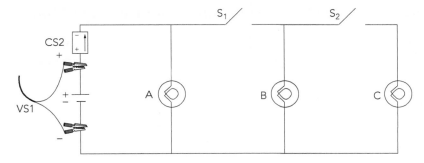

Figure 6-9: Circuit to examine voltage across a not-so-fresh battery as the current through the battery increases.

4. Measure the voltage across the battery and the current flowing through the battery with both switches open, with S1 closed and with both switches closed (Table 6-1).

Table 6-1

	Voltage (V)	Current (A)
Both switches open		
S1 closed		
Both switches closed		

Question 2-7: Did the voltage across this not-so-fresh battery remain constant as the current flowing through the battery increased? If not, how did it change?

Batteries are sources of potential energy for the charges flowing through them. They also have an *internal resistance* that increases in size as they wear out. The equivalent circuit of a battery with internal resistance is shown in Figure 6-10.

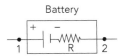

Figure 6-10: Equivalent circuit of a battery with internal resistance.

Question 2-8: Are your measurements for voltage and current in Table 6-1 consistent with the equivalent circuit in Figure 6-10? Explain.

If you have time, work on the following extension.

Extension 2-3: Applying Your Current and Voltage Models

Let's return to a more complex circuit using what we now know about voltage and current. In Lab 5, Investigation 3, you explored the circuit shown in Figure 6-11.

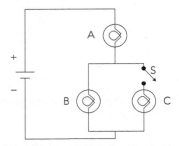

Figure 6-11: Circuit equivalent to Figure 5-9a when the switch is open, and to Figure 5-9b when the switch is closed.

You were previously asked to rank the brightness of bulbs A, B, and C after the switch was closed. The question now is, *what happens to the brightness of bulb B when the switch is closed*? Does it increase, decrease, or remain the same?

Prediction E2-3: Based on the current and voltage models you have developed, *carefully* predict what will happen to the current flowing through bulb B (and therefore its brightness) when bulb C is added in parallel to it. Will it increase, decrease, or remain the same? Explain the reasons for your answer.

Connect the circuit in Figure 6-11, and make observations. Describe what happens to the brightness of bulb B when the switch is closed.

Question E2-9: Did your observations agree with your prediction? If not, use the current and voltage models to explain your observations.

INVESTIGATION 3: OHM'S LAW

What is the relationship between current and potential difference? You have already seen on several occasions that there is only a potential difference across a bulb or resistor when there is a current through the circuit element. The next question is *how does the potential difference depend on the current*?

To explore this, you will need the following:

- computer-based laboratory system
- current and voltage sensors
- *RealTime Physics Electricity and Magnetism* experiment configuration files
- variable regulated DC power supply (up to 3 V and 0.5 amps)
- 6 alligator clip leads
- 10-ohm resistor
- #14 bulb in a socket

Examine the circuit shown in Figure 6-12. A variable DC power supply is like a variable battery. When you turn the dial, you change the voltage (potential difference) between its terminals. Therefore, this circuit allows you to measure the current flowing through the resistor when different voltages are across it.

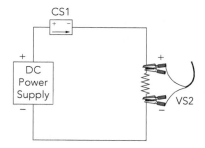

Figure 6-12: Circuit with a variable power supply to explore the relationship between current and potential difference for a resistor.

Prediction 3-1: What will happen to the current flowing *through the resistor* as you turn the dial on the power supply and increase the applied voltage from zero?

Prediction 3-2: What will happen to the potential difference *across the resistor* as the current flowing through it increases from zero?

Prediction 3-3: What will be the mathematical relationship between the *voltage across the resistor* and the *current flowing through the resistor*?

Activity 3-1: Current and Potential Difference for a Resistor

1. Open the experiment file called **Ohm's Law (L06A3-1).**

2. **Calibrate** the current and voltage sensors or **load the calibration,** if this has not already been done. **Zero** both sensors with the red and black alligator clips of the voltage sensor connected together, and nothing connected to the current sensor.

3. Connect the circuit in Figure 6-12. Note that the current sensor is connected to measure the current flowing through the resistor, and is plugged into Channel 1, and the voltage sensor is connected to measure the potential difference across the resistor, and is plugged into Channel 2.

    Your instructor will show you how to operate the power supply.

4. **Begin graphing** current and voltage with the power supply set to zero voltage, and graph as you turn the dial and increase the voltage *slowly* to about 3 V.

Warning: Do not exceed 3 V!

Question 3-1: What happened to the current flowing in the circuit as the power supply voltage was increased? Did this agree with your prediction?

Question 3-2: How did the potential difference across the resistor change as the current flowing through the resistor changed? Did this agree with your prediction?

5. You can display axes for voltage vs. *current* on the bottom graph by **adjusting the horizontal axis** to read **Current 1**. The axes should now be as shown below.

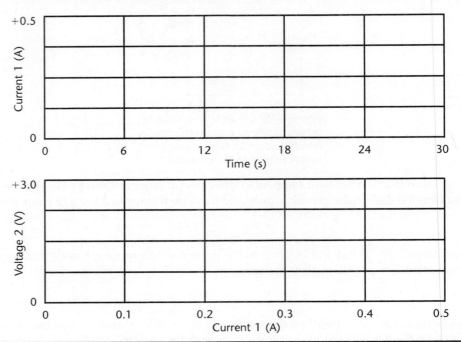

> **Reminder:** We are interested in the nature of the mathematical relationship between the voltage across the resistor and the current flowing through the resistor. This can be determined from the graph by drawing a smooth curve that fits the plotted data points. Some definitions of possible mathematical relationships are shown below. In these examples, y might be the voltage reading and x the current.
>
>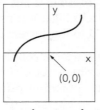
>
> y is a function of x which increases as x increases.
>
>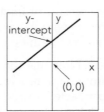
>
> y is a *linear* function of x which increases as x increases according to the mathematical relationship y = mx + b, where b is a constant called the *y*-intercept.
>
>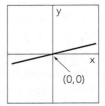
>
> y is *proportional* to x. This is a special case of a linear relationship where y = mx, and b, the *y-intercept*, is zero.
>
> These graphs show the differences between these three types of mathematical relationship. y can increase as x increases, and the relationship doesn't have to be *linear* or *proportional*.
>
> *Proportionality* refers only to the special linear relationship where the *y*-intercept is zero, as shown in the example graph on the right.

6. Use the **fit routine** in the software to see if the relationship between voltage and current for a resistor is a proportional one.

7. Sketch your graphs on the axes above, or **print** them and affix them over the axes.

Question 3-3: In words, what is the mathematical relationship between potential difference and current for a resistor? Explain based on your graphs.

Comment: The relationship between potential difference and current that you have observed for a resistor is known as Ohm's law. The definition of resistance is $R = V/I$.

If potential difference is measured in volts and current is measured in amperes, then the unit of resistance is the ohm, which is usually represented by the Greek letter "omega" (Ω).

Question 3-4: Based on your graph, what can you say about the value of R for a resistor—is it constant, or does it change as the current flowing through the resistor changes? Explain.

Question 3-5: From the slope of your graph, what is the experimentally determined value of the resistance of your resistor in ohms? Does this agree with the value written on the resistor?

Note: Many circuit elements do not obey Ohm's law. The definition for resistance is still the same ($R = V/I$), but the resistance changes as the current changes. Circuit elements that follow Ohm's law—like resistors—are said to be *ohmic*. Circuit elements that have different resistances when the current changes (such as diodes), and so do not obey Ohm's law are said to be *non-ohmic*.

In the last activity you explored the relationship between the potential difference across a resistor and the current flowing through the resistor. It is a proportional relationship. Instead of a resistor, in the following extension you will explore the relationship between current and potential difference for a light bulb.

Extension 3-2: Relationship Between Current and Potential Difference for a Light Bulb

1. Replace the 10-Ω resistor by the light bulb (Figure 6-13).

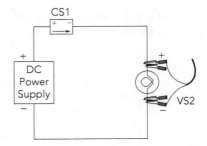

Figure 6-13: Circuit with a variable power supply to explore the quantitative relationship between the current and potential difference for a light bulb.

Prediction E3-4: What do you predict will happen to the brightness of the bulb as you turn the dial on the power supply and increase the voltage from zero? Explain.

Prediction E3-5: What will be the mathematical relationship between the *voltage across the bulb* and the *current flowing through the bulb*?

2. Prepare to graph current vs. time with sensor 1 and voltage vs. time with sensor 2. (**Adjust the horizontal axis** on the bottom graph back to **time**.)

3. **Begin graphing** with the power supply set to zero voltage, and graph current and voltage as you turn the dial and increase the voltage *slowly* to about 3 V.

Warning: Do not exceed 3 V since this may burn out the bulb!

Question E3-6: What happened to the brightness of the bulb as the power supply voltage was increased? Did this agree with your prediction?

Question E3-7: How is the brightness of the bulb related to the potential difference across the bulb? To the current flowing through the bulb?

4. You can again display a graph of potential difference vs. current by **adjusting the horizontal axis** on the bottom graph to **Current 1** as before.

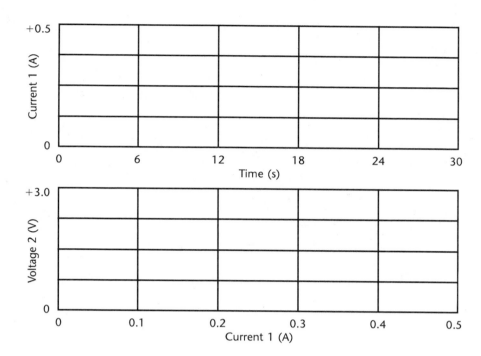

Question E3-8: Compare your graph of voltage vs. current for the bulb to that for the resistor in the previous activity. In what ways are they similar, and in what ways are they different?

5. Sketch the graphs, or **print** and affix them over the axes.

Question E3-9: Does the relationship between voltage and current for a light bulb appear to be a proportional one. Explain.

Question E3-10: Based on your graph of voltage vs. current for a bulb, what can you say about the value of R for a bulb—is it constant, or does it change as the current flowing through the bulb changes? Explain.

Question E3-11: Is a light bulb an ohmic device? Explain.

HOMEWORK FOR LAB 6
VOLTAGE IN SIMPLE DC CIRCUITS AND OHM'S LAW

1. In the circuit below, the battery maintains a constant potential difference between its terminals, points 1 and 2 (i.e., the internal resistance of the battery is considered negligible).

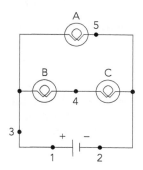

The three light bulbs, A, B, and C are identical.

a. How do the brightnesses of the three bulbs compare to each other? Explain your reasoning.

b. What happens to the brightness of each of the three bulbs when bulb A is unscrewed and removed from its socket? Explain your reasoning.

c. When A is unscrewed, what happens to the current through points 3, 4, and 5? Explain your reasoning.

d. Bulb A is screwed back in. What happens to the brightness of each of the three bulbs when bulb C is unscrewed and removed from its socket? Explain your reasoning.

e. When C is unscrewed, what happens to the current flowing through points 3, 4, and 5? Explain your reasoning.

2. For each of the questions A–E below, a wire is connected from the battery terminal at point 1 to point 4.

a. What happens to the brightness of each of the three bulbs? Explain.

b. What happens to the current flowing through point 3? Explain.

c. What happens to the potential difference across bulb B? Explain.

d. What happens to the potential difference across bulb C? Explain.

e. What happens to the potential difference between points 1 and 5? Explain your reasoning.

3. The wire described in (2) is removed. What happens to the brightness of each of the three bulbs and to the current flowing through point 2 if a wire is connected from the battery terminal at point 2 to the socket terminal at point 5?

4. The circuit is returned to its original state. A fourth bulb (D) is connected in parallel with bulb B (*not in parallel with B and C*).

   a. Sketch the bulb in the circuit.
   b. What happens to the brightness of each of the three bulbs?

   c. What happens to the current flowing through point 3?

   d. What happens to the potential difference between points 3 and 4?

   e. What happens to the potential difference between points 4 and 2?

5. State Ohm's law in words. For what type of circuit elements does it correctly describe the behavior?

6. Does a light bulb have a constant resistance? Explain. Is a light bulb *ohmic*?

7. Does a resistor have a constant resistance? Explain. Is a resistor *ohmic*?

8. Draw diagrams for a 75-Ω and a 100-Ω resistor connected in series and connected in parallel:

               Series   Parallel

9. In the following circuits, tell which resistors are connected in series, which are connected in parallel, and which are neither in series nor parallel.

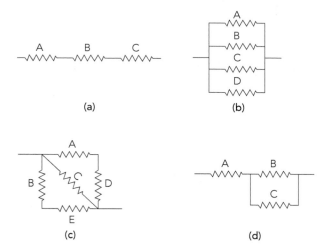

Pre-Lab Preparation Sheet for
Lab 7—Kirchhoff's Circuit Rules

(Due at the beginning of lab)

Directions:
Read over Lab 7 and then answer the following questions about the procedures.

1. What is a multimeter? What quantities can you measure with one?

2. In order to measure the potential difference across a resistor, should a multimeter be connected in series or in parallel with the resistor?

3. A resistor has four colored stripes in the following order: orange, violet, red, and silver. What is its rated resistance and the tolerance?

4. How will you measure resistances to determine a rule for calculating the effective resistance of one or more resistors wired in parallel.

5. State Kirchhoff's Loop Rule.

6. State Kirchhoff's Junction Rule.

7. Which of Kirchhoff's rules is based on charge conservation?

LAB 7:
KIRCHHOFF'S CIRCUIT RULES

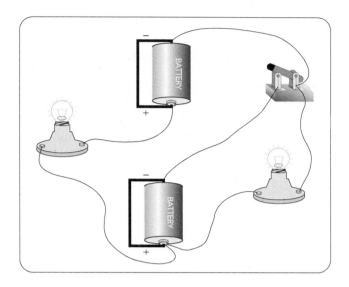

The Energizer© keeps on going, and going, and . . . .

—Eveready Battery Company, Inc.

OBJECTIVES

- To learn how multimeters are designed so that they don't modify the currents and voltages being measured.

- To learn to measure resistance with a multimeter.

- To develop a method for calculating the equivalent resistance of resistors connected in series.

- To develop a method for calculating the equivalent resistance of resistors connected in parallel.

- To understand Kirchhoff's circuit rules and use them to determine the currents that flow in various parts of DC circuits.

OVERVIEW

In the last few labs, you have examined simple circuits with bulbs or resistors connected in series and parallel. The emphasis has been on learning about the concepts of current, voltage, and resistance in fairly simple DC circuits.

In this lab you will look at circuits more quantitatively. Up until now you have been using computer-based sensors to measure currents and voltages. A *multimeter* is a device with the capability of measuring current and voltage. Some multimeters can also be used to measure resistance. In Investigation 1, you will learn how to use a multimeter to measure current, voltage, and resistance.

In Investigation 2, you will look at circuits with resistors connected in both series and parallel and discover the rules for finding the equivalent values of networks of resistors wired in series and parallel.

Sometimes circuit elements are connected with multiple batteries in more complicated ways than simply in series or parallel. The rules for series and parallel addition of resistances are not adequate to determine the currents flowing in such circuits. In Investigation 3 of this lab, you will learn about Kirchhoff's circuit rules that are generally applicable to all types of circuits.

INVESTIGATION 1: MEASURING CURRENT, VOLTAGE, AND RESISTANCE

Resistance, voltage, and current are fundamental electrical quantities that characterize all electric circuits. The multimeters available to you can be used to measure these quantities. All you need to do is choose the correct dial setting, connect the wire leads to the correct terminals on the meter, and connect the meter correctly in the circuit. Figure 7-1 shows a simplified diagram of a multimeter.

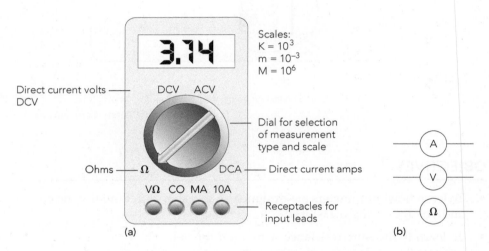

Figure 7-1: (a) Multimeter with voltage, current, and resistance modes, and (b) symbols that will be used to represent a multimeter when it is used as an ammeter, voltmeter, or ohmmeter, respectively.

A current sensor and a multimeter used to measure current are both connected in a circuit in the same way. Likewise, a voltage sensor and a multimeter used to measure voltage are both connected in a circuit in the same way. The next two activities will remind you how to connect them. The activities will also show you that when meters are connected correctly, they don't change the currents or voltages being measured.
You will need:

- digital multimeter
- 2 very fresh, akaline 1.5-V D cell batteries with holders
- 6 alligator clip leads
- 2 #14 bulbs and sockets

Activity 1-1: Measuring Current with a Multimeter

Figure 7-2 shows two possible ways that you might connect a multimeter to measure the current flowing through bulb 1.

Prediction 1-1: Which of the diagrams in Figure 7-2, (b) or (c), shows the correct way to connect a multimeter to measure the current through bulb 1? Explain why it should be connected this way. [**Hint:** In which case is the current flowing through the multimeter the same as that flowing through bulb 1?]

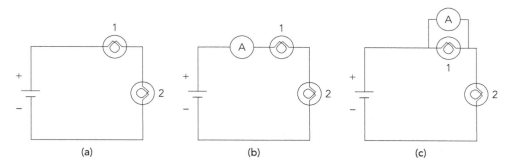

Figure 7-2: (a) A circuit with two light bulbs and a battery, and two possible *but not necessarily desirable* ways to connect a multimeter to measure current: (b) in series with bulb 1, and (c) in parallel with bulb 1.

1. Set up the basic circuit in Figure 7-2a. Use two batteries in series to make a 3-V battery. Observe the brightness of the bulbs.

2. Set the multimeter to measure current and connect it as shown in Figure 7-2b. Was the brightness of the bulbs significantly affected?

3. Now connect the meter as in Figure 7-2c. Was the brightness of the bulbs significantly affected?

Question 1-1: If the multimeter is connected correctly to measure current, it should measure the current flowing through bulb 1 without significantly affecting the current flowing through the bulb. Which circuit in Figure 7-2 shows the correct way to connect a multimeter? Explain based on your observations. Why is it connected in this way?

Question 1-2: Does the multimeter appear to behave as if it is a large resistor or a small resistor? Explain based on your observations. Why is it designed in this way?

Activity 1-2: Measuring Voltage with a Multimeter

Figure 7-3 shows two possible ways that you might connect a multimeter to measure the potential difference across bulb 1.

Prediction 1-2: Which of the diagrams in Figure 7-3 shows the correct way to connect a multimeter to measure the voltage across bulb 1? Explain why it should be connected this way.

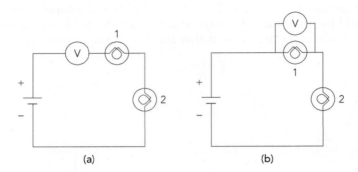

Figure 7-3: Two possible *but not necessarily desirable* ways to connect a multimeter to measure voltage: (a) in series with bulb 1, and (b) in parallel with bulb 1.

1. Set the meter to measure voltage and connect it as in Figure 7-3a. Was the brightness of the bulbs significantly affected?

2. Now connect the meter as in Figure 7-3b. Was the brightness significantly affected?

Question 1-3: If the multimeter is connected correctly, it should measure the voltage across bulb 1 without significantly affecting the current flowing through the bulb. Which circuit in Figure 7-3 shows the correct way to connect the multimeter? Explain based on your observations. Why is it connected in this way? [**Hint:** In which case is the voltage across the multimeter the same as that across bulb 1?]

Question 1-4: Does the multimeter behave as if it is a large resistor or a small resistor? Explain based on your observations. Why is it designed in this way?

You just observed that even multimeters have some resistance. Now you will investigate how to measure resistance with a multimeter.

In earlier labs, you observed that the resistance of a light bulb increases when the current through it causes the temperature of the filament to rise. To make the behavior of circuits more consistent, it is desirable to have circuit elements with resistances that do not change. For that reason, circuit elements known as *resistors* have been invented. The resistance of a well-designed resistor doesn't vary with the amount of current passing through it (or with temperature), and resistors are inexpensive to manufacture.

The most common resistors contain a form of carbon known as graphite suspended in a hard glue binder. It is usually surrounded by a plastic case with a color code painted on it. Figure 7-4 depicts a carbon resistor cut down the middle.

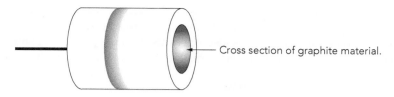

Figure 7-4: A cutaway view of a carbon resistor.

Figure 7-5 shows a carbon resistor with colored bands that tell you the value of the resistance and the tolerance (guaranteed accuracy) of this value.

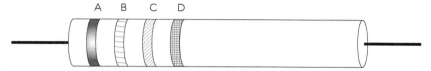

Figure 7-5: A carbon resistor with color bands.

The first two stripes indicate the two digits of the resistance value. The third stripe indicates the power of ten multiplier, and the fourth stripe signifies the resistor's tolerance. The key in Table 7-1 shows the corresponding values.

Table 7-1: The resistor code

Bands A, B, C	violet = 7
black = 0	gray = 8
brown = 1	white = 9
red = 2	
orange = 3	Band D
yellow = 4	none = ±20%
green = 5	silver = ±10%
blue = 6	gold = ±5%

As an example, look at the resistor in Figure 7-6. Its two digits are 1 and 2 and the multiplier is 10^3, so its value is 12×10^3, or 12,000 Ω. The tolerance is ±20%, so the value might actually be as large as 14,400 Ω or as small as 9600 Ω.

Figure 7-6: An example of a color-coded carbon resistor. The resistance of this resistor is 12×10^3 Ω ± 20%.

The appropriate way to connect the multimeter to measure resistance is shown in Figure 7-7. When the multimeter is in its ohmmeter mode, it connects a known voltage across the resistor and measures the current through the resistor. Then resistance is calculated by the meter from $R = V/I$.

Note: Resistors must be isolated (disconnected from the circuit) before their resistances can be measured. This also prevents damage to the multimeter that may occur if a voltage is connected across its terminals while in the resistance mode.

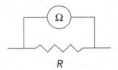

Figure 7-7: Connection of an ohmmeter to measure resistance.

In the next activity, you will use the multimeter to measure the resistance of several resistors. You will need

- several color-coded resistors
- 2 digital multimeters
- 6-V battery

Activity 1-3: Reading Resistor Codes and Measuring Resistance

1. Choose several resistors and read their codes. Record the resistances and the tolerances in the first two columns of Table 7-2.

Table 7-2

R from code (Ω)	Tolerance from code	Measured R (Ω)	Measured V (V)	Measured I (A)	R = V/I (Ω)

There are two ways to determine resistance with the multimeter. One is to use the resistance mode and measure the resistance directly. The second is to connect the resistor to a battery and then measure the voltage across the resistor and the current through the resistor. R can then be calculated from Ohm's law, R = V/I. You will use both of these methods to measure the resistances of the resistors you selected.

2. Measure the resistance of each of your resistors directly using the multimeter and enter the values in Table 7-2 under "Measured R."

3. *Be sure to reset the multimeter to measure current and voltage, respectively, or you will burn out the fuse in the meter.* Connect each resistor to the battery and simultaneously measure the voltage across the resistor and the current through the resistor with the multimeters. Record V and I in the appropriate columns of Table 7-2 and calculate the resistance from these values.

Question 1-5: How do the values of your resistors measured with the resistance mode of the multimeter compare to the values indicated by the code? Assuming that your measured values are correct, are the values indicated by the code correct within the stated tolerance?

Question 1-6: How do the resistance values found from the voltage across the resistor and the current through it compare to the values measured with the resistance mode of the multimeter? Do you conclude that the resistance mode is reliable? Explain.

INVESTIGATION 2: SERIES AND PARALLEL COMBINATIONS OF RESISTORS

Several resistors can be wired in *series* or in *parallel* as shown in Figure 7-8.

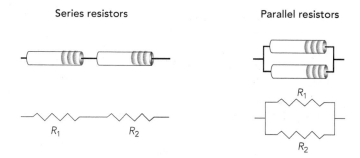

Figure 7-8: Resistors wired in series and in parallel.

Prediction 2-1: Do you think that two identical resistors wired in series will have a total resistance that is greater, the same as, or less than the individual resistance of one of them?

To do some exploration of equivalent resistance of different resistors wired in combination you will need the following:

- 3 51-Ω resistors
- a 22-Ω and a 75-Ω resistor
- digital multimeter
- 6 alligator clip leads

Activity 2-1: Equivalent Resistance of Resistors Connected in Series

1. Measure the actual values of the 51-Ω resistors with the multimeter. Record their values below:

   R_1 _____ Ω R_2 _____ Ω R_3 _____ Ω

2. Connect R_1 and R_2 in series. Measure the resistance of this series combination with the multimeter.

   Resistance of R_1 and R_2 in series: _____ Ω

3. Now connect R_1, R_2, and R_3 in series and measure the resistance of the combination.

   Resistance of R_1, R_2, and R_3 in series: _____ Ω

Question 2-1: How did your measurements compare to your prediction?

Question 2-2: Based on your measurements, state a rule for finding the equivalent resistance of several resistors connected in series. If the resistors have resistances R_1, R_2, and R_3, write a mathematical equation for the equivalent resistance, R_{eq}, when these are connected in series. Explain how your measurements support this rule.

Question 2-3: Does your rule agree with your observations in Lab 5 that the current through two identical resistors connected in series is half the current through a single resistor connected to the same battery? Explain.

Prediction 2-2: Do you think that two identical resistors wired in parallel will have a total resistance that is greater, the same as, or less than the individual resistance of one of them?

Test your prediction.

Activity 2-2: Equivalent Resistance of Resistors Connected in Parallel

1. Use your three 51-Ω resistors again. Connect R_1 and R_2 in parallel. Measure the resistance of this parallel combination with the multimeter.

   Resistance of R_1 and R_2 in parallel: _____ Ω

2. Now connect R_1, R_2, and R_3 in parallel, and measure the resistance of the combination.

   Resistance of R_1, R_2, and R_3 in parallel: _____ Ω

Question 2-4: How did your measurements compare with your prediction?

Question 2-5: Based on your measurements, is the equivalent resistance, R_{eq}, consistent with the following mathematical relationship?

$$\frac{1}{R_{eq}} = \frac{1}{R_1} + \frac{1}{R_2} + \frac{1}{R_3}$$

Show the calculations you used to check the validity of this equation.

Question 2-6: Does the relationship in Question 2-5 agree with your observations in Lab 5 that the current through a battery connected to two identical resistors connected in parallel is twice the current through the battery when connected to a single resistor? Explain.

Extension 2-3: Other Combinations of Resistors in Series and Parallel

1. Measure the values of the 22-Ω and 75-Ω resistors with the multimeter. Record these values below:

   R_4 _____ Ω R_5 _____ Ω

2. Connect R_1, R_4, and R_5 in series. Measure the resistance of this series combination.

   Resistance of R_1, R_4, and R_5 in series:_____ Ω

Question E2-7: Use your rule in Question 2-2 to calculate the equivalent resistance of these three resistors in series. Show your calculation. How does this value compare to the measured resistance of the series combination?

3. Connect R_1, R_4, and R_5 in parallel. Measure the resistance of this parallel combination.

   Resistance of the combination:_____ Ω

Question E2-8: Use the rule in Question 2-5 to calculate the equivalent resistance of these three resistors. Show your calculation. How does this value compare to the measured resistance of the parallel combination?

Now that you know the basic rules for calculating the equivalent resistance for series and parallel connections of resistors, you can tackle the question of how to find the equivalent resistance for a complex network of resistors. The trick is to be able to calculate the equivalent resistance of each segment of the complex network and use that in calculations of the next segment.

For example, in the network shown in Figure 7-9, there are two resistance values, R_1 and R_2. A series of simplifications is shown in the diagram below.

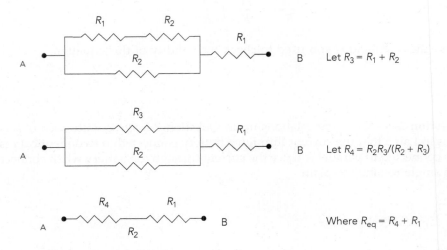

Figure 7-9: A sample resistor network.

To complete the following extension about equivalent resistance you will need:

- 3 22-Ω resistors
- 3 51-Ω resistors
- digital multimeter

Extension 2-4: The Equivalent Resistance for a Network

1. Use the color-coded value for your smaller resistors for R_1 and the color-coded value for your larger resistors for R_2. List these values below.

   R_1:_____ R_2:_____

2. Calculate the equivalent resistance between points A and B for the network shown below. Show your calculations on a step-by-step basis.

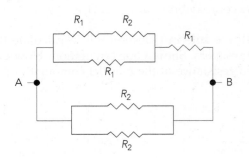

3. Set up this network of resistors and check your calculation by measuring the equivalent resistance directly with the multimeter.

Question E2-9: How did your measured value for the equivalent resistance agree with the calculated value? Could any disagreement be explained by the tolerances in the resistor values? Explain in detail.

INVESTIGATION 3: KIRCHHOFF'S CIRCUIT RULES

Consider a circuit that has many components wired together in a complex array. Suppose you want to calculate the currents in various branches of this circuit. The rules for combining resistors that you examined in Investigation 2 are very convenient in circuits made up only of resistors that are connected in series or in parallel. But, while it may be possible in some cases to simplify parts of a circuit with the series and parallel rules, complete simplification to an equivalent resistance is often impossible. This is especially true when additional components such as more than one battery are included. The application of Kirchhoff's circuit rules can help you to understand the most complex circuits.

Kirchhoff's circuit rules are based on two conservation laws that apply to circuits: conservation of charge and conservation of energy. Before summarizing these rules, we need to define the terms *junction* and *branch* in a circuit. Figure 7-10 illustrates the definitions of these two terms for an arbitrary circuit. As shown in Figure 7-10a, a junction in a circuit is a place where two or more wires are connected together. As shown in Figure 7-10b, a branch is a portion of the circuit in which the current is the same through all of the circuit elements. (That is, the circuit elements in a branch are all connected in series with each other.)

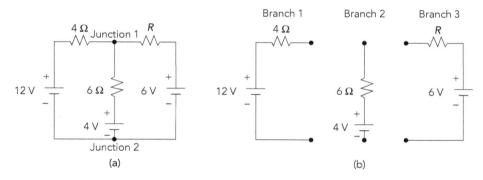

Figure 7-10: An arbitrary circuit used to illustrate junctions and branches.

Kirchhoff's rules can be summarized as follows:

1. *Junction Rule (based on charge conservation):* The sum of all the currents entering any junction of the circuit must equal the sum of the currents leaving.

2. *Loop Rule (based on energy conservation):* Around any closed loop in a circuit, the sum of all changes in potential (emfs and potential drops across resistors and other circuit elements) must equal zero.

These rules can be applied to a circuit using the following steps:

1. Assign a current symbol to each branch of the circuit, and label the current in each branch I_1, I_2, I_3, etc.

2. *Arbitrarily* assign a direction to each current. (The direction chosen for the current in each branch is arbitrary. If you choose the right direction, when you solve the equations, the current will come out positive. If you choose the wrong direction, the current will come out negative, indicating that its direction is actually opposite to the one you chose.) Remember that the current is always the same everywhere in a branch, and the current out of a battery is always the same as the current into a battery.

3. Apply the *Loop Rule* to each of the loops.

   a. Let the potential drop (voltage) across each resistor be the negative of the product of the resistance and the net current through the resistor. (However, make the sign positive if you are traversing a resistor in the direction opposite that of the current).

   b. Assign a positive potential difference when the loop traverses from the − to the + terminal of a battery. (If you are going across a battery in the opposite direction, assign a negative potential difference to the trip across the battery terminals.)

4. Find each of the junctions and apply the *Junction Rule* to it. You can write currents leaving the junction on one side of the equation and currents coming into the junction on the other side of the equation.

To illustrate the application of the rules let's consider the circuit in Figure 7-11.

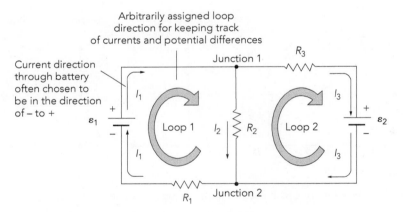

Figure 7-11: A complex circuit in which loops 1 and 2 share the resistor R_2.

Question 3-1: Why aren't the resistors R_1 and R_2 in series? Why aren't they in parallel?

In Figure 7-11 the directions for the loops through the circuits and for the three currents are assigned arbitrarily. That is, other assignments would work. If we assume that the internal resistances of the batteries are negligible, then by applying the *Loop Rule* we find that

Loop 1 $$+\varepsilon_1 - I_2R_2 - I_1R_1 = 0 \qquad (1)$$

Loop 2 $$-\varepsilon_2 + I_2R_2 - I_3R_3 = 0 \qquad (2)$$

By applying the *Junction Rule* to junction 1 or 2, we find that

$$I_1 = I_2 + I_3 \qquad (3)$$

(current into junction = current out of junction)

It may trouble you that the current directions and directions that the loops are traversed have been chosen arbitrarily. You can explore this assertion by changing these choices and analyzing the circuit again. (Alternate assignments if applied consistently will yield the same final results!) To do the next activity you'll need the following:

- 2 resistors (rated values of 39 and 75 Ω)
- digital multimeter (to measure resistance)
- 6-V battery
- very fresh, alkaline 1.5-V D cell battery and holder

Activity 3-1: Applying the Loop and Junction Rules Several Times

Figure 7-12 shows the same circuit as in Figure 7-11 with different arbitrary directions for the loops and current through R_2.

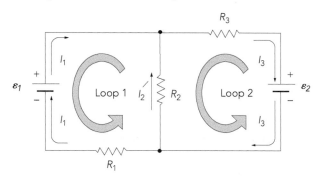

Figure 7-12: The same complex circuit as in Figure 7-11 with the current through R_2 chosen arbitrarily in the opposite direction, and the loops traversed in the counterclockwise direction instead of clockwise.

1. Use the loop and junction rules to write down the three equations for Figure 7-12 that correspond to Equations (1), (2), and (3) derived above for Figure 7-11.

Question 3-2: Show that if you make the substitution $I_2' = -I_2$, then the three equations you just derived can be rearranged algebraically so they are *exactly the same* as Equations (1), (2), and (3).

2. Measure the actual values of the two fixed resistors of 75 and 39 Ω and the two battery voltages with your multimeter. List the results below.

Measured voltage (emf) of the 6-V battery ε_1:_____

Measured voltage (emf) of the 1.5-V battery ε_2:_____

Measured resistance of the 75-Ω resistor R_1:_____

Measured resistance of the 39-Ω resistor R_3:_____

3. Carefully rewrite Equations (1), (2), and (3) with the appropriate *measured* (*not rated*) values for emf and resistances substituted into them. Use 100 Ω for the value of R_2 in your equations. You will be setting a variable resistor to that value soon.

4. Solve these three equations for the three unknown currents, I_1, I_2, and I_3 in amps. Show your calculations in the space below.

Question 3-3: Do your currents actually satisfy the equations? Use direct substitution to find your answer.

Now you can verify that the application of Kirchhoff's rules actually works for this circuit.

In addition to the materials from the previous activity, you will need:

- 0 to 200-Ω variable resistor (potentiometer)
- 6 alligator clip leads

Activity 3-2: Testing Kirchhoff's Rules with a Real Circuit

1. Before wiring the circuit, use the resistance mode of the multimeter to measure the resistance between the center wire on the variable resistor and one of the other wires. What happens to the resistance reading as you rotate the dial on the variable resistor clockwise? Counterclockwise?

2. Set the variable resistor so that there is 100 Ω between the center wire and one of the other wires. Was it difficult to do? If so, explain why.

3. Wire up the circuit pictured in Figure 7-11 using the 0 to 200-Ω variable resistor set at 100 Ω as R_2. Spread the wires and circuit elements out on the table so that the circuit looks as much like Figure 7-10 as possible.

4. Use the multimeter to measure the current in each branch of the circuit (see Note below), and enter your data in Table 7-3. Compare the measured values to those calculated in Activity 3-1 by computing the percent difference in each case.

> **Note:** The most accurate and easiest way to measure the currents with the digital multimeter is to measure the voltage across a resistor of known value, and then use Ohm's law to calculate I from V and R.

Table 7-3

	R measured w/ multimeter (Ω)	V measured w/ multimeter (V)	Measured $I = V/R$ (amps)	Theoretical I (amps) (from Activity 3-1)	% Difference
R_1					
R_2					
R_3					

Question 3-4: Use your measured current values to verify the junction rule at the two junctions in the circuit.

Question 3-5: Use your measured voltage values to verify the loop rule in loop 1 labeled in Figure 7-11.

Question 3-6: How well do your measured currents agree with the theoretical values you calculated in Activity 3-1? Are they within a few percent, or do they differ by more than this?

If you have additional time, do the following extension.

Extension 3-3: How Do Changes Affect the Currents?

In Activities 3-1 and 3-2 you analyzed the circuit in Figure 7-11 with Kirchhoff's Loop Rule and Kirchhoff's Junction Rule. Now consider the case where resistor R_2 is removed from the circuit.

1. Use the space to the right to draw a picture of the modified circuit (with R_2 removed).

Question E3-7: Will the analysis of the modified circuit with Kirchhoff's rules now be more or less complex than it was for the circuit in Figure 7-11? That is, will you need to generate more or fewer equations with Kirchhoff's Loop Rule? With Kirchhoff's Junction Rule? Why?

Question E3-8: What is the effective value of R_2 in the modified circuit? Think about this question very carefully and explain.

Prediction E3-1: Predict the values of the voltages across R_1 and R_3 for the modified circuit. Show your calculations clearly in the space below.

Test your prediction. Remove R_2 and measure I_1 and I_3 from V_1, V_3, R_1, and R_3 as before. Record the new values below.

V_1:_____ V_3:_____

I_1:_____ I_3:_____

Question E3-9: Did your observations agree with your predictions? Explain.

HOMEWORK FOR LAB 7
KIRCHHOFF'S CIRCUIT RULES

1. Find the equivalent resistance of the following network. (All resistances are in ohms.) Show your work below.

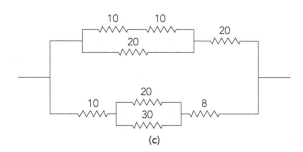

(c)

2. Show on the circuit diagram in Question 1 how you would connect a multimeter to measure the current through the 8-Ω resistor.

   a. Explain why the multimeter is connected in this way.

   b. What characteristic of a good multimeter allows you to connect it in the way you indicated without appreciably affecting the current through the 8-Ω resistor?

3. Show on the circuit diagram in Question 1 how you would connect a multimeter to measure the voltage across the 8-Ω resistor.

   a. Explain why the multimeter is connected in this way.

   b. What design feature of a good multimeter allows you to connect it in the way indicated without appreciably affecting the voltage across the 8-Ω resistor?

4. A battery with emf 12 V and internal resistance 1 Ω is pictured in the thick-lined box in the circuit on the right. A is the positive terminal of the battery and B is the negative terminal.

   a. What is the potential difference across the battery's terminals when the switch S is open, as shown?

b. What is the potential difference across the battery's terminals when the switch S is closed?

5. Determine the values and tolerances of resistors with the following color codes:

   a. red-red-brown-gold

   b. violet-gray-blue

   c. Could the value of a resistor marked as in (a) actually be as large as 240 Ω? Explain.

6. Find the current through each of the resistors in the circuit on the right. Give both the magnitude and direction of current flow. Show all of your work below.

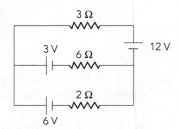

PRE-LAB PREPARATION SHEET FOR
LAB 8—INTRODUCTION TO CAPACITORS
AND RC CIRCUITS

(Due at the beginning of lab)

Directions:
Read over Lab 8 and then answer the following questions about the procedures.

1. Predict the change in capacitance of a parallel plate capacitor as the area of the plates is increased.

2. Predict the change in capacitance of a parallel plate capacitor as the separation between the plates is increased.

3. Briefly describe what observations you will make in Activity 1-2 to test one of these two predictions.

4. If you have two identical capacitors, what do you predict will happen to the capacitance if they are connected in parallel?

5. Briefly describe what observations you will make in Activity 2-1 to test your prediction.

6. Sketch below the complete circuits in Figure 8-5 with the switch in position 1, and with the switch in position 2.

7. What devices will you use to measure how the voltage across a capacitor decreases over time in an RC circuit?

LAB 8:
INTRODUCTION TO CAPACITORS AND RC CIRCUITS

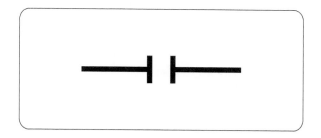

I get a real charge out of capacitors.

—P. W. Laws

OBJECTIVES

- To define capacitance and to learn how to measure it with a digital multimeter.

- To discover how the capacitance of conducting parallel plates is related to the area of the plates and their separation.

- To explore and apply the rules for finding the equivalent capacitance of several capacitors connected in parallel and of several capacitors connected in series.

- To discover the effect of connecting a capacitor in a circuit in series with a resistor or bulb and a voltage source.

- To discover how the charge on a capacitor and the current through it change with time in a series circuit containing a capacitor, a resistor, and a voltage source.

OVERVIEW

Capacitors are widely used in electronic circuits where it is important to store charge and/or energy or to trigger a timed electrical event. For example, circuits with capacitors are designed to do such diverse things as setting the flashing rate of Christmas lights, selecting what station a radio picks up, and storing the electrical energy needed to fire an electronic flash unit. Any pair of conductors that can be charged electrically so that one conductor has excess positive charge and the other conductor has an equal amount of excess negative charge on it is called a capacitor.

A capacitor can be made up of two differently shaped blobs of metal or it can have any number of regular symmetric shapes, such as one hollow metal sphere inside another, or a metal rod inside a hollow metal cylinder (Figure 8-1).

Amorphous
capacitor (blobs)
with air as an
insulator

Cylindrical capacitor with air
as an insulator

Parallel-plate capacitor
with paper and air
as insulators

Figure 8-1: Some different capacitor geometries.

The easiest type of capacitor to analyze is the parallel-plate capacitor. We will focus exclusively on studying the properties of parallel-plate capacitors because they are easy to construct and their behavior can be predicted using simple mathematical calculations and basic physical reasoning.

Although many of the most interesting properties of capacitors show up in the operation of AC (alternating current) circuits (where current is first in one direction and then in the other), we will limit our present study to the behavior of capacitors in DC (direct current) circuits like those you have been constructing in the last couple of labs.

The circuit symbol for a capacitor is simply a pair of parallel lines, as shown in Figure 8-2. Note that it is similar to the symbol for a battery, except that both lines are the same length for the capacitor.

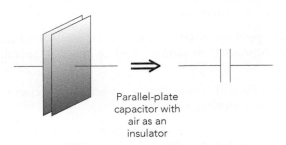

Parallel-plate
capacitor with
air as an
insulator

Figure 8-2: The circuit diagram symbol for a capacitor.

INVESTIGATION 1: CAPACITANCE, AREA, AND SEPARATION

The usual method for transferring equal and opposite charges to the plates of a capacitor is to use a battery or power supply to produce a potential difference between the two conductors. Electrons will then flow from one conductor (leaving a net positive charge) to the other (making its net charge negative) until the potential difference produced between the two conductors is equal to that of the battery (see Figure 8-3).

In general, the amount of charge needed to produce a potential difference equal to that of the battery will depend on the size, shape, and location of the conductors relative to each other as well as the properties of the material between the conductors. The capacitance of a given capacitor is defined as the ratio of the magnitude of the excess charge, q (on either one of the conductors), to the voltage (potential difference), V, applied across the two conductors.

Thus,

$$C = q/V$$

Capacitance is defined as a measure of the magnitude of the net or excess charge on either one of the conductors per unit potential difference. So a capacitor that can store more charge at a given voltage has a larger capacitance.

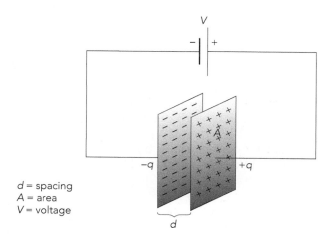

d = spacing
A = area
V = voltage

Figure 8-3: A parallel-plate capacitor with a voltage V across it.

You can draw on some of your experiences with electrostatics to think about what might happen to a parallel-plate capacitor when it is hooked to a battery, as depicted in Figure 8-3. This thinking can give you an intuitive feeling for the meaning of capacitance. For a fixed voltage from a battery, the net charge found on either plate is proportional to the capacitance of the pair of conductors and the applied voltage, $q = CV$.

Activity 1-1: Predicting the Dependence of Capacitance on Area and Separation

Consider two identical parallel metal plates of area A separated by a distance d. The space between the plates is filled with a nonconducting material (air, for instance). Suppose each plate is connected to one of the terminals of a battery.

Question 1-1: What type of excess charge will build up on the metal plate that is attached to the negative terminal of the battery? What type of excess charge will build up on the plate that is connected to the positive terminal of the battery? Explain.

Question 1-2: Can the excess positive charges on one plate of a charged parallel-plate capacitor exert forces on the excess negative charges on the other plate? Explain.

Question 1-3: Consider two plates of area A that are separated by a distance d as shown in Fig. 8-3. If the area, A, of the plates were increased (with the spacing and potential difference between the plates held constant) what do you think would happen to the amount of excess charge on each of the plates? Explain your reasoning. How will this affect the capacitance of the capacitor? [**Hint:** Do the electric field and potential difference between the plates depend on the total charge on each plate or on the charge per unit area?]

Question 1-4: If the battery is then disconnected what do you think would happen to the potential difference between the plates if the separation d were decreased with the excess charge on each plate held constant? Explain. [**Hint:** What happens to the electric field between the plates as d decreases while the excess charge is kept constant by disconnecting the capacitor from the battery? What happens to the potential difference across the plates as d is made smaller after the capacitor is disconnected from the battery?]

The unit of capacitance is the farad, F, named after Michael Faraday. One farad is equal to one coulomb/volt. As you should be able to demonstrate to yourself shortly, the farad is a very large capacitance. Thus, actual capacitances are often expressed in smaller units with alternative notation as shown below:

microfarad: $1~\mu F = 10^{-6}$ F

nanofarad: $1~nF = 1000~\mu\mu F = 10^{-9}$ F

picofarad: $1~pF = 1~\mu\mu F = 1~UUF = 10^{-12}$ F

(Note that M, m, μ, and U when written on a capacitor all stand for a multiplier of 10^{-6}.)

Several types of capacitors are typically used in electronic circuits, including disk capacitors, foil capacitors, and electrolytic capacitors. You should examine some typical capacitors. Your instructor will supply you with:

- a collection of old capacitors to look at

In the next few activities you will construct a parallel-plate capacitor, measure its capacitance and examine how it depends on spacing and plate area. You will use the following items:

- 2 sheets of aluminum foil 20 cm × 20 cm
- pages in a "fat" textbook or phone book
- one or several bricks or other massive objects
- digital multimeter with a capacitance mode and clip leads
- ruler with a centimeter scale
- vernier calipers or a micrometer (optional)

- computer-based laboratory or data analysis software
- *RealTime Physics Electricity and Magnetism* experiment configuration files

You can construct a parallel-plate capacitor out of two rectangular sheets of aluminum foil separated by pieces of paper. Pages in a textbook or phone book work well as the separators for the foil sheets. You can slip the two foil sheets between any number of pages, and weight the book down with something heavy like some bricks or another, heavy book. The digital multimeter can be used to measure capacitance.

Activity 1-2: Measuring How Capacitance Depends on Area or on Separation

Be sure that you understand how to set the meter and how to connect a capacitor to it. Devise a way to measure how the capacitance depends on *either* the foil area or the separation between foils. Of course, you must keep the other variable (separation or area) constant.

When you measure the capacitance of your "parallel plates," be sure that the aluminum foil pieces are pressed together as uniformly as possible, and that they don't make electrical contact with each other.

If you hold the separation constant, record its value in Table 8-1. This may be measured in "pages," or the vernier caliper or micrometer may be used to translate this into meters. The area may be varied by using different size sheets of aluminum foil.

Alternatively, if you hold the area constant and vary separation, then record the dimensions of the foil so you will be able to calculate the area and enter it in Table 8-1.

1. Take at least five data points in either case. Record your data in Table 8-1.

Table 8-1

Separation (m)	Length (m)	Width (m)	Area (m^2)	Capacitance (nF)

Question 1-5: Describe how you measured all of the quantities in Table 8-1. Explain how you varied the separation or the area.

2. After you have collected all of your data, open the experiment file called **Dependence of C (L08A1-2)**. **Enter** your data for capacitance and either separation or area from Table 8-1 into the table in the software. Graph capacitance vs. either separation or area. **Print** the graph and affix it to these pages.

3. If your graph looks like a straight line, use the **fit routine** in the software to find its equation. If not, you should try other functional relationships until you find the best fit. **Print** and affix any other graphs.

Question 1-6: What mathematical relationship best describes the dependence of capacitance on plate separation or plate area. How do the results compare with your prediction based on physical reasoning?

Question 1-7: What difficulties did you encounter in making accurate measurements?

The actual mathematical expression for the capacitance of a parallel-plate capacitor of plate area A and plate separation d (Figure 8-3) is derived in your textbook. The result when there is vacuum (or just a low-density gas) between the plates is

$$C = \varepsilon_0 A/d$$

where $\varepsilon_0 = 8.85 \times 10^{-12} \, C^2/Nm^2$.

Question 1-8: Do your predictions and/or observations on the variation of capacitance with plate area and separation agree qualitatively with this result? Explain.

Question 1-9: Use one set of your actual areas and separations that corresponds to measurements you recorded in Table 8-1 to calculate a value of C using this equation. Show your calculations. How does the calculated value of C compare with your measured value? What might be wrong with the theoretical model for a capacitor in describing the behavior of your capacitor in this activity?

Question 1-10: In theory, what length and width in miles would big square foil sheets separated by a distance of 1 mm with wax paper have to be on each side for you to construct a 1-F capacitor? Show your calculations. Assume that wax paper has the same electrical properties as air. [**Hint:** Miles are not meters! In fact, 1000 m = 1 km = 0.62 mile.]

$L =$ _____ miles

INVESTIGATION 2: CAPACITORS IN SERIES AND PARALLEL

Capacitors come in all sizes, shapes, and colors. Take a look at the array of actual capacitors. You can measure their capacitances with the multimeter. You can also connect them in various series and parallel combinations and measure the equivalent capacitances of these combinations.

The definitions of series and parallel wiring are the same as for other circuit elements such as resistors as shown in Figure 8-4.

In a series connection, there is only one path for the charge. Whatever charge is placed on one of the capacitors must also be transferred to the other(s). In a parallel connection, the two terminals of each capacitor are connected directly to the terminals of the other(s). Each capacitor defines a branch, so that the total charge transferred to the capacitor combination is divided among the different capacitors.

To examine the equivalent capacitance of two identical capacitors connected in parallel or series, you'll need:

- 2 different capacitors (each about 0.1 μF)
- multimeter with a capacitance mode
- 6 alligator clip leads

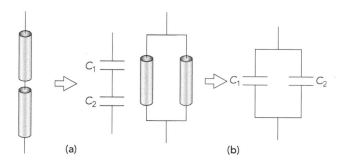

Figure 8-4: Capacitors wired (a) in series and (b) in parallel.

Activity 2-1: Equivalent Capacitance for Parallel Connection of Capacitors

Prediction 2-1: Use direct physical reasoning to predict the equivalent capacitance of a pair of capacitors with capacitance C_1 and C_2 wired in parallel. Explain your reasoning below. [**Hint:** What is the effective area of two parallel-plate capacitors wired in parallel? Does the effective separation between plates change when the capacitors are connected in parallel?]

1. Measure the capacitance of each capacitor with the multimeter. Don't forget to include units!

   C_1:_____ C_2:_____

2. Connect the two capacitors in parallel and measure the equivalent capacitance of the parallel combination.

   C_{eq}:_____

Question 2-1: From your measurements, figure out a general equation for the equivalent capacitance of a parallel network in terms of C_1 and C_2. Explain how you reached your conclusion.

Question 2-2: How did your equation agree with your prediction? Explain.

Prediction 2-2: Use direct physical reasoning to predict the equivalent capacitance of a pair of capacitors wired in series. Explain your reasoning. [**Hint:** If you connect two capacitors in series, what will happen to the charge along the conductor between them? What will the effective separation of the "plates" be? Will the effective area change?]

3. Connect the same two capacitors in series and measure the equivalent capacitance of the series combination.

    C_{eq}:————

Question 2-3: Are your measurements, consistent with the following equation for combining capacitors connected in series into an equivalent capacitance C_{eq}?

$$\frac{1}{C_{eq}} = \frac{1}{C_1} + \frac{1}{C_2}$$

Show the calculations you used to reach your conclusion.

Question 2-4: Is the equation given in Question 2-3 consistent with your prediction? Explain.

INVESTIGATION 3: CHARGE BUILDUP AND DECAY IN CAPACITORS

Capacitors can be connected with other circuit elements. When they are connected in circuits with resistors, some interesting things happen. In this investigation you will explore what happens to the voltage across a capacitor when it is placed in series with a resistor in a direct current circuit. From your observations, you should be able to devise qualitative and quantitative explanations of what is happening.

For the activities in this investigation you will need:

- computer-based laboratory system
- 2 current and 2 voltage sensors
- *RealTime Physics Electricity and Magnetism* experiment configuration files

- 6-V battery
- #133 flashlight bulb and socket
- 2 capacitors with very large capacitance (about 25,000 μF)
- 6 alligator clip wires
- single-pole–double-throw switch
- 2 22-Ω resistors

You can first use a bulb in series with one of the amazing ultra-large capacitors. These will allow you to see what happens. Later on, to get a more quantitative result, the bulb will be replaced by a resistor, with a resistance that doesn't change with temperature.

Activity 3-1: Observations with a Capacitor, Battery, and Bulb

1. Set up the circuit shown in Figure 8-5. (If you are using a "polar" capacitor with + and − signs on its inputs, be sure that its positive and negative terminals are connected correctly.)

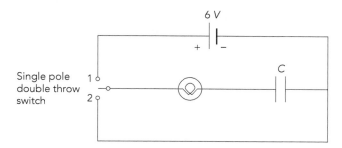

Figure 8-5: Circuit to examine the charging and discharging of a capacitor through a light bulb.

Question 3-1: Sketch the complete circuit for current when the switch is in position 1, and when it is in position 2.

Position 1   *Position 2*

Prediction 3-1: What do you predict will happen to the brightness of the bulb when you move the switch to position 1 for awhile?

2. Start with the switch in position 2 for at least 30 s. Then move the switch to position 1 and observe the brightness of the bulb for awhile.

Question 3-2: Draw a sketch on the axes below of the *approximate* brightness of the bulb as a function of time for the above case where you move the switch to position 1 after it has been in position 2 for a long time. Let $t = 0$ be the time when the switch was moved to position 1.

Prediction 3-2: What do you predict will happen to the brightness of the bulb after you move the switch back to position 2?

3. Now move the switch back to position 2. Observe the brightness of the bulb for awhile.

Question 3-3: Draw a sketch on the axes below of the *approximate* brightness of the bulb as a function of time when it is placed across a charged capacitor *without the battery present,* i.e., when the switch is moved to position 2 after being in position 1 for a long time. Let $t = 0$ when the switch is moved to position 2.

Question 3-4: Can you explain why the bulb behaves in this way? Is there charge on the capacitor after the switch is in position 1 for awhile? What happens to this charge when the switch is moved back to position 2?

4. Open the experiment file called **Capacitor Decay (L08A3-1),** and display the axes that follow.

5. Connect the two sensors to the interface, and **calibrate** them or **load the calibration. Zero** the sensors with nothing attached to them.

6. Connect the sensors in the circuit as in Figure 8-6 to measure the current through the light bulb and the potential difference across the capacitor.

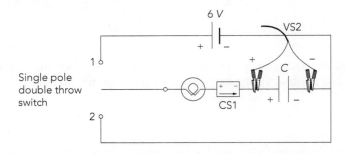

Figure 8-6: Current and voltage sensors connected in Figure 8-5.

REALTIME PHYSICS: ELECTRICITY AND MAGNETISM

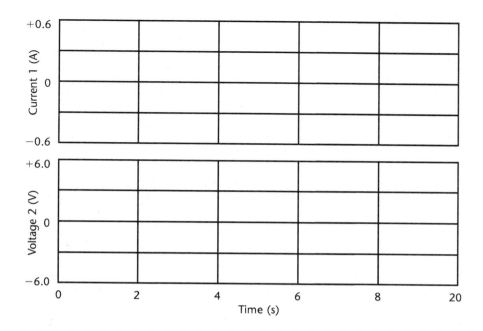

7. Move the switch to position 2. **Begin graphing.** When the graph begins tracing, move the switch to position 1. When the current and voltage stop changing, move the switch back to position 2.

8. Sketch the graphs on the axes above, or **print** them and affix them over the axes.

9. Indicate on the graphs the times when the switch was moved from position 2 to position 1, and when it was moved back to position 2 again.

Question 3-5: Does the actual behavior over time observed on the current graph agree with your sketches in Questions 3-2 and 3-3? Do any features of the graphs surprise you? Explain.

Question 3-6: Based on the graph of potential difference across the capacitor, can you explain why the bulb lights when the switch is moved from position 1 to position 2 (when the bulb is connected to the capacitor with no battery in the circuit). Also explain the way the brightness of the bulb changes with time.

As you have seen before, a bulb does not have a constant resistance. Instead, its resistance is temperature-dependent and goes up when it is heated by the current through it. For more quantitative studies of the behavior of a circuit with resistance and capacitance, you should replace the bulb with a 22-Ω resistor.

Activity 3-2: The Rise and Decay of Voltage in an RC Circuit

1. Replace the light bulb in your circuit (Figure 8-6) with a 22-Ω resistor. Move the switch to position 2. **Begin graphing.** When the graph begins to trace, move the switch to position 1. When current and voltage stop changing, move the switch back to position 2.

2. Sketch your graphs on the axes that follow, or **print** them and affix them over the axes.

3. Indicate on the graphs the times when the switch was moved from position 2 to position 1, and when it was moved back to position 2 again.

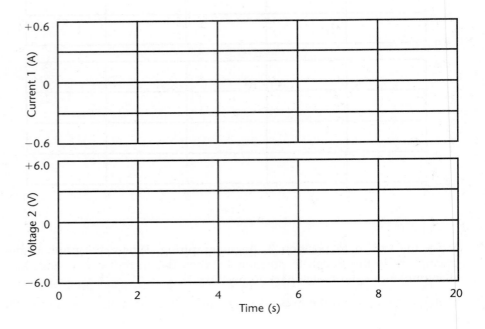

Question 3-7: Do the graphs for the resistor appear similar to those for the bulb? Are there any significant differences?

4. Use the **analysis feature** of the software to determine from your graph the *time constant* (the time for the voltage across the capacitor to decay to 37% of its initial value—after the switch is moved to position 2). Record your data below.

Initial voltage:_____   Time:_____

37% of initial voltage:_____   Time:_____

Time constant:_____

Comment: If you made careful measurements of V vs. t for a capacitor C discharging through a resistor R, you should have obtained what is known as an *exponential decay curve*. This curve has exactly the same mathematical form as the cooling curve you may have encountered in the study of heat and temperature.

Mathematical reasoning based on Ohm's law as well as the definitions of current and capacitance can be used to show that the following equation represents the voltage $v(t)$ across the capacitor as a function of time:

$$v(t) = V_0 \, e^{-t/RC}$$

In this equation, V_0 is the initial potential difference across the capacitor. (Note that V_0 is not necessarily the voltage of the battery.)

Question 3-8: Do the curves you measured for the decay of the potential difference across the capacitor in series with a resistor have the shape described by an exponential decay? How do you know?

Question 3-9: Use the exponential function to calculate the *time constant* for your capacitor–resistor combination. [**Hint:** What time t in the function would make the value for $v(t)$ be just $0.37V_0$? (This is just V_0/e, where e is the base of natural logarithms.)] Use your values for R and C. Show all work. Does this value agree with your measured value?

Does this mathematical function describe the data you collected? If you have more time, do the following extension to examine this question.

Extension 3-3: Does the Observed Decay Curve Fit Theory?

1. Use the **fit routine** in the software to see if the voltage decay curve can be fit by an exponential function. Be sure to **select** and fit only that portion of the voltage graph where the voltage is decaying (decreasing) that is, up to the time when the voltage has *just reached* its minimum value.

2. **Print** the graph and affix it over the axes above, and record the equation of the function in Question 3-10.

Question E3-10: Did the exponential function fit your data well? Is the decay of voltage across the capacitor an exponential decay?

Question E3-11: Find the value of RC from the exponent in the exponential function used to fit the data. Compare this value to the one calculated from the resistance and capacitance. Do they agree? (Remember that the resistance and capacitance values are each known only to about $\pm 10\%$.)

If you have more time, do the following extension.

Extension 3-4 Decay with a Larger Resistance and Capacitance

Prediction E3-1: What do you predict will happen if a larger resistance is connected in series with the capacitor? How will this affect the time constant?

Test your prediction.

1. Increase the resistance to twice its value. (Should you connect two 22-Ω resistors in series or in parallel?)

2. Start with the switch in position 2 and graph as before. Again measure the time constant from your graph. Record your data below.

Initial voltage:_____   Time:_____

37% of initial voltage:_____   Time:_____

Time constant:_____

Question E3-12: What happened to the time constant? Did this agree with your prediction?

Prediction E3-2: What will happen if the capacitance of the capacitor is larger? How will this affect the time constant?

Test your prediction.

3. Increase the capacitance to twice its value. (Should you connect two capacitors in series or in parallel?)

4. Start with the switch in position 2 and graph as before. Again measure the time constant from the graph. Record your data below.

Initial voltage:_____   Time:_____

37% of initial voltage:_____   Time:_____

Time constant:_____

Question E3-13: What happened to the time constant? Did this agree with your prediction?

Name_____ Date_____ Partners_____

HOMEWORK FOR LAB 8
INTRODUCTION TO CAPACITORS
AND RC CIRCUITS

1. Explain in terms of the charge, electric field, and potential difference how the capacitance of a parallel-plate capacitor depends on the area and separation of the plates in the equation $C = \varepsilon_0 A/d$.

2. If a 1.5-V battery is connected to a 250-μF capacitor, how much excess charge is there on each of the capacitor plates?

3. For the circuit on the right with two capacitors of different capacitance in series, indicate whether the statements below are TRUE or FALSE, and for each false statement, write a correct one.

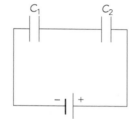

   a. Both capacitors have the same amount of charge on their plates.

   b. The voltages across the capacitors are the same.

   c. The sum of the voltages on the two capacitors equals the voltage of the battery.

4. Find the equivalent capacitance of each of the following combinations of capacitors. (All capacitances are in μF.)

   a.
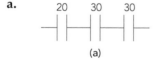

   b.
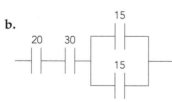

   c.
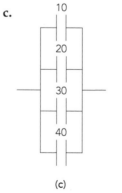

5. In the circuit on the right, the capacitor is initially uncharged.

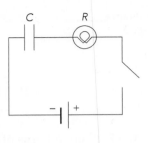

   a. Describe what is observed when the switch is closed.

   b. How would your observations be changed if the capacitor were twice as large?

   c. How would your observations be changed if the bulb had half as much resistance?

6. Sketch a graph of the current as a function of time in the circuit in Question 5 after the switch is closed. Also sketch a graph of the voltage across the capacitor as a function of time.

7. In the circuit on the right, the capacitor is initially charged. If it has capacitance 0.023 F, and the resistor has resistance 47 Ω, how long after the switch is closed will it take for the voltage on the capacitor to fall to 37% of its initial value? Show your calculations.

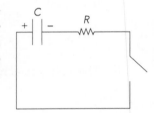

PRE-LAB PREPARATION SHEET FOR
LAB 9—MAGNETISM

(Due at the beginning of lab)

Directions:
Read over Lab 9 and then answer the following questions about the procedures.

1. Describe briefly what types of observations you will make in Activities 1-1 and 1-2 to examine the interactions of magnetic poles with various objects and with each other.

2. How will you determine in Activity 1-4 which pole of your magnet is North and which is South?

3. How will you determine the direction of magnetic field lines around a bar magnet in Activity 1-6?

4. How will you measure the strength of the magnetic field around a bar magnet in Extension 1-7.

5. What does the Lorentz force law describe?

LAB 9:
MAGNETISM

To you alone . . . who seek knowledge, not from books only, but also from things themselves, do I address these magnetic principles and this new sort of philosophy. If any disagree with my opinion, let them at least take note of the experiments . . . and employ them to better use if they are able.

—William Gilbert (1544–1603)

OBJECTIVES

• To learn about the properties of permanent magnets and the forces they exert on each other.

• To understand how a magnetic field can be represented by field lines.

• To explore a magnetic field law that is the equivalent of Gauss' law for electric fields.

• To understand how materials can be magnetized.

• To understand the Lorentz force equation, a mathematical representation of the force on a charged particle moving in a magnetic field.

• To observe the force exerted on a current carrying wire in a magnetic field.

OVERVIEW

As a child, you probably played with small magnets and used compasses. Magnets exert forces on each other. The Earth is a magnet. Magnets are used in electrical devices such as meters, motors, and loudspeakers. Magnetic materials are used in cassette tapes and computer disks. Large electromagnets consisting of current-carrying wires wrapped around pieces of iron are used to pick up automobiles in junkyards.

Although electricity and magnetism are not the same thing, the fascinating thing is that, from a theoretical perspective, these two phenomena that appear different in many ways, are closely related. For example, electric currents that are caused by electric fields can cause magnetic effects. Permanent magnets can exert forces on current-carrying wires and vice versa. Electric currents can produce magnetic fields, and changing magnetic fields can, in turn, produce electric fields.

In contrast to our earlier study of electrostatics, which focused on the forces between resting charges, the study of magnetism is at heart the study of the forces acting between moving charges. In this lab, you will explore some of these relationships.

INVESTIGATION 1: MAGNETIC FORCES AND FIELDS

Permanent Magnets

The attraction of iron to a magnet is so familiar that we seldom realize that most of us know little more than the ancients about how the attraction occurs. Let us begin our exploration of magnetism by playing carefully and critically with some permanent magnets and observing what happens. For the activities involving permanent magnets you will need:

- 2 rod-shaped permanent magnets (with "like" ends marked blue or red)
- aluminum rod (the same size and shape as the magnets)
- 5 tiny compasses
- paper clip
- pen
- pencil
- 2 plastic objects
- 2 strings, 10 cm long
- Scotch Magic© tape
- rod stand and right-angle clamp (not made of iron or steel)
- 2 aluminum rods

Magnetic Interactions with Permanent Magnets and Other Objects

Permanent magnets can interact with each other as well as with other objects. Let's explore the forces exerted by one magnet on another and then make qualitative observations of forces that a magnet can exert on other objects.

Prediction 1-1: Assuming that two rod-shaped magnets are identical, do you expect like ends to attract or repel each other? What do you predict if you bring unlike ends together?

Test your predictions.

Activity 1-1: Permanent Magnet Poles

1. Fiddle with the two permanent magnets. Explore the interactions between the different ends of the two magnets.

Question 1-1: Do the like ends attract or repel each other? What about unlike ends? How do the rules of attraction and repulsion compare to those for electrical charges of like or different sign? Are the rules the same or different?

Note: One of the ends of the magnet is called the *north pole* and the other the *south pole*. You will explore why shortly.

Question 1-2: Each pole represents a different type of magnetic "charge." Can you find a magnet with just a north pole or just a south pole? Can you find unlike electrical charges separately? Discuss the difference between electrical charges and magnetic poles.

Prediction 1-2: List at least four objects, one of which is an aluminum rod. *Predict* what will happen if you bring each object near one pole of a magnet and then near the other pole of that magnet.

Pred.	Object	Pole 1	Pole 2
1			
2			
3			
4			

Test your predictions.

Activity 1-2: Permanent Magnet Interactions with Objects

1. Observe what happens when you bring the various objects close to each of the poles of the magnet, and summarize your results in the table below.

Obs.	Object	Pole 1	Pole 2
1			
2			
3			
4			

Question 1-3: How do your predictions and observations compare? Be specific.

Magnetic Orientations

Next, let's explore how a suspended magnet orients itself when it is placed close to another identical magnet. What happens when the suspended magnet is placed far away from any other magnets? For this activity you will need to tie a string *tightly* around the center of a magnet and put a small piece of Scotch Magic© tape under the string as shown in the figure below. Set up a rod stand, rods, and clamp to suspend the magnet a few centimeters above the table. **Note:** The stand, rods, clamp and table should be non-magnetic (e.g., aluminum). Be sure that iron objects are far away from the magnets.

Prediction 1-3: Based on your previous observations, do you expect the suspended magnet to align itself parallel or anti-parallel to a stationary magnet? Why?

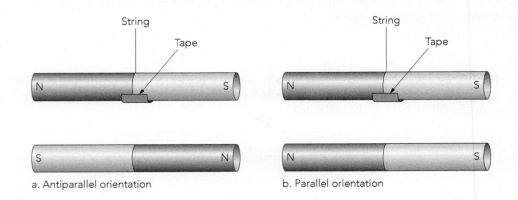

a. Antiparallel orientation   b. Parallel orientation

Activity 1-3: Magnet Orientation

1. Place another identical magnet on the table underneath the suspended magnet as shown in the figure above.

2. Observe the suspended magnet, and in the figure above circle its actual orientation relative to the fixed magnet. Cross out the orientation that you don't observe.

3. Next, suspend both of the magnets at some distance away from each other so they don't interact.

Question 1-4: Do the non-interacting suspended magnets appear to be oriented relative to each other? If yes, what is the direction of their orientations? Cite evidence for and against orientation.

Question 1-5: If the suspended magnet still appears to orient itself, what might be underneath the room? Does the orientation happen outdoors? If it's convenient, take a suspended magnet outside. What might be under the ground?

4. Repeat (3) with two of the small compasses.

Question 1-6: What does a compass needle probably consist of? Explain.

By now, you should have discovered that the Earth behaves as if it has a rod-shaped magnet embedded in it. The Earth's magnet also has two poles—a north pole and a south pole, just like any other magnet.

Comment: Want to know something crazy!? The North Pole on the Earth attracts the north pole on the magnet. *So, the Earth's geographic North Pole is actually a magnetic south pole.* This is just as bad as the way Ben Franklin defined the signs of charges so that, once they were discovered, electrons—the usual current-carrying charges—turned out to be negative!

Activity 1-4: Determining Magnetic Poles

1. Make observations to determine which poles of your magnets are north—the red poles or the blue ones.

Question 1-7: Are the north poles of your large permanent magnets red or blue? Describe how you determined this.

2. Make observations to determine which poles of your compasses are north. (**Warning:** Sometimes cheap little compasses get magnetized the "wrong" way. Check a *group* of little compasses.)

Question 1-8: Which pole of your compass is the north pole? Describe how you determined this. (On the average, which poles are the north poles?)

Suppose you have a magnet made up of several small disks stuck together. How does its behavior compare to that of the rod magnets you have been using?

Prediction 1-4: How do you predict the pieces will behave if you break them into two pieces, four pieces, and so on?

To test your prediction you will need:

- 4 ceramic disk shaped "refrigerator" magnets
- small compass
- rod-shaped magnet

Activity 1-5: Magnetic Disks Together and Apart

1. Place the four disks in a stack and compare the behavior of the resulting magnet with that of the rod-shaped magnet.

Question 1-9: Does it behave like the unbroken rod-shaped magnet? Why or why not? Does it have North and South poles?

2. Pull the stack apart into two stacks with two disks in each stack. Compare the behavior of one of the resulting magnets with that of the rod-shaped magnet.

Question 1-10: Does it behave like the unbroken rod-shaped magnet? Why or why not? Does it have North and South poles?

3. Pull the magnets apart completely. Compare the behavior of one of the resulting magnets with that of the rod-shaped magnet.

Question 1-11: Does it behave like the unbroken rod-shaped magnet? Why or why not? Does it have North and South poles?

Question 1-12: Do the poles of an individual disk magnet behave the same way when the disk is in the center of a stack? Explain!

Question 1-13: What do you think would happen if you cut one of the individual magnets in half along its axis? Does it seem possible to split a magnet into one with just a north pole and another with just a south pole?

Magnetic "Field" Lines Around a Rod-shaped Magnet

In preparation for defining a quantity called magnetic field, which is analogous to, but not the same as, an electric field, you should explore the alignments of the small compass at various places near the larger rod-shaped magnets. This will then allow you to postulate the existence of magnetic flux and of a mathematical law for magnetic flux not unlike Gauss' law for electrical flux.

The direction of the magnetic field at a point in space is defined as the direction that a small magnet (e.g., compass needle) would point (which way the North pole would point) if it was placed at that point. By determining the orientation of a compass at different points around a rod-shaped magnet, you have effectively determined what the magnetic field around the rod-shaped magnet looks like. Of course, it is difficult to determine the orientation at all points around a rod-shaped magnet by using something as large as a compass needle. However, there is a very nice method for determining the magnetic field around a rod-shaped magnet that uses the fact that iron becomes magnetized in a magnetic field. For the observations that follow you will need the following:

- rod-shaped magnet with N and S poles marked on it
- small compass
- iron filings
- paper and cardboard

Activity 1-6: Field Directions Around a Rod-shaped Magnet and Gauss' Law for Magnetic Fields

1. Use a small compass to map the magnetic field directions in the space surrounding a rod-shaped magnet. Denote the direction of the field with arrows. Sketch lots and lots of arrows in the space below.

2. Place two pieces of cardboard (or two notebooks) right up next to the magnet on both sides. The cardboard (or notebooks) should be about as thick as the magnet. Next, place a sheet of white paper on top of the cardboard so that the magnet is approximately under the center of the paper.

3. Start sprinkling iron filings on top of the white paper starting at the center (where the magnet is) and then moving outward. Keep sprinkling until you can see a nice pattern. The iron filings become oriented along the direction of the magnetic field lines.

4. Using the results of both observations—with the compass and with the iron filings—make a sketch in the space below of the field lines around *and inside* the magnet. **Note:** Assume that the lines are continuous across the boundaries between the magnetic material and the surrounding air.

5. Mark lines with directional arrows pointing in the direction that the north pole of your compass points.

Question 1-14: Were the patterns observed with the compass needle and with the iron filings consistent with each other? Explain.

Question 1-15: Try to explain what was happening to the iron filings to cause the pattern that you saw.

6. Pretend you are in a two-dimensional world. (Flatland again!) Draw several closed loops in the magnetic field line diagram above. Let one loop enclose no magnetic pole, another loop enclose one of the poles, and another enclose both poles (the entire magnet).

Question 1-16: Note the direction of the arrows you drew around the bar magnet shown above. Assuming that each line coming into a loop is negative and each line coming out is positive, what is the net number of magnetic field lines coming in and out of the loop in each case?

Question 1-17: Now we come to the magnetic equivalent to Gauss's law describing the net magnetic flux, Φ^{mag}, coming out of a closed three-dimensional surface. Can you guess what Φ^{mag} is equal to? **Hint:** Use the answer to Question 1-16 to predict what the net magnetic flux is equal to.

$\Phi^{mag} =$

Question 1-18: Recall from Lab 2 that Gauss' law states that electric flux is proportional to the charge enclosed by a closed Gaussian surface. If the magnetic flux through a closed surface is equal to zero in *all situations*, is it possible to have a net magnetic charge enclosed within the surface? **Hint:** According to your observations in Activities 1-5 and 1-6, are north and south poles ever separate?

Extension 1-7: Magnetic Field Strength Around a Rod-shaped Magnet

Now you will use a magnetic field sensor to measure the strength of the field around the rod-shaped magnet. You will need

- computer-based laboratory system
- *RealTime Physics Electricity and Magnetism* experiment configuration files
- magnetic field sensor

The sensor should always be aligned so that the field lines are perpendicular to the face of the device. (Or, in other words, the readings of the sensor will be largest when its face is perpendicular to the field lines.)

Correct orientation of the magnetic field sensor

1. Open the experiment configuration file **Measuring Mag. Field (L09E1-7)** to set up the software to use the magnetic field sensor.

2. Probe the field strength around the magnet. Start perpendicular to and against one face and move the probe away slowly. Find the approximate distances from the end of the magnet where the field falls to 1/2 and 1/4 of its value at the face.

	North Pole	South Pole
Field Magnitude	Distance from Face	Distance from Face
B_{face}	0	0
½ B_{face}		
¼ B_{face}		

3. Repeat for the other end of the magnet.

Question E1-19: Describe your measurements. How do the measurements differ for the two ends of the magnet?

4. Now probe other areas around the magnet, and indicate on the diagram below where the field is strongest (SS), strong (S) and where it is weak (W) and weakest (WW). (Use your knowledge of field directions from Activity 1-6 to align the field sensor correctly.)

INVESTIGATION 2: THE BEHAVIOR OF MAGNETIC MATERIALS[1]

In the last investigation, you determined how two magnets interact with each other. But what exactly *is* a magnet? This is not a simple question. In this investigation, you will take a closer look at the behavior of magnetic materials in the

[1] These activities were adapted from *Explorations in Physics* (Hoboken, NJ, Wiley, 2003).

presence of a magnet. This process will allow you to describe what a magnet is, or at least how one goes about making one.

You will need:

- 2 strong rod-shaped magnets
- several paper-clips
- Petri dish

Activity 2-1: Magnets and Paperclips

1. Play around with the magnet and paperclips. Beginning with two paper-clips on the table, try to figure out a way to pick up both paperclips without using your hands, such that only one of the paperclips is touching the magnet.

Question 2-1: How far can you get the paperclip from the magnet and still lift it? Explain your procedure below. If it helps, draw a simple picture.

Question 2-2: Is the paperclip (the one not touching the magnet) being attracted to the magnet or to the other paperclip? Explain briefly and be sure to support your answer with evidence. Again, feel free to draw a picture if it helps you explain your answer.

A magnet with two paperclips attached.

2. Now try the following experiment. Hold the magnet so that one paperclip hangs from one of the poles of the magnet. Then, gently pick up a second paperclip onto the end of the first paperclip, as shown in the figure. Then, holding onto the upper paperclip with one hand, *gently* detach the magnet from the paperclip.

3. Now try using the upper paperclip to pick up a different paperclip.

Question 2-3: What do you notice? What appears to have happened to the upper paperclip?

4. Now repeat the experiment with a slight variation. Begin by picking up two paperclips with the magnet as before. Again, hold the upper paperclip and gently detach the magnet. Then, flip the magnet over and slowly bring the opposite pole toward the top of the upper paperclip *without touching it*.

Question 2-4: Describe what you observe and explain what the magnet seems to be doing to the paperclip. You might want to try this experiment a few times to make certain your results are reproducible.

A paperclip is magnetized between two magnets.

Comment: The last activity is very interesting. It appears that a paperclip can be turned into a weak magnet simply by coming into contact with a magnet. This phenomenon is called *magnetization*. In fact, further experiments show that there does not need to be any physical contact between the magnet and the paperclip for this to happen. The paperclip only needs to be held close to the magnet. Of course, to verify that the paperclip has been turned into a little magnet requires us to observe that the paperclip indeed has two poles like a regular magnet. This is the goal of the next activity.

Activity 2-2: The Paperclip Magnet

1. Take a paperclip and place it between opposite poles of two magnets as shown in the figure to the left. *Remember which end of the paperclip is in contact with which pole of the magnet.*

2. While one person holds the magnets and paperclip in this configuration (it should stay in place if you lay it down on the table), someone else can go and get a small Petri dish filled with water.

3. Attempt to float the paperclip on the surface of the water as follows. After the paperclip has been between the magnets for a few minutes, take it out and hold it between your thumb and forefinger on the long sides of the paperclip. Then, lower your hand toward the dish until your fingers come into contact with the water. When the paperclip is very nearly touching the water (within a few millimeters), gently open your fingers and let the paperclip drop very gently onto the surface of the water. If you are careful, your paperclip should be floating on the surface of the water. If it sinks, you will need to take it out and try it again with a dry paperclip.

Prediction 2-1: How will the floating paperclip behave when you bring the N pole of a magnet nearby. Your prediction should be very specific.

4. Place one of the magnets on the table far away from the Petri dish and arrange it so that the N pole is closest to the Petri dish.

5. Slowly bring the magnet closer to the Petri dish until you notice an interaction. **Note:** The magnet should not be brought too close to the Petri dish. Always keep it at least a couple of inches away.

Question 2-5: Describe your observations. Was your prediction correct?

Question 2-6: Check to see if your paperclip exhibits both attractive and repulsive behavior, and if so, explain which end of the paperclip is a N pole and which end is a S pole. Be sure to support your answers with evidence.

6. Now try magnetizing another paperclip with the same orientation as you did in (1).

Prediction 2-2: What will you observe when you bring this magnetized paperclip near the floating paperclip. Be specific and make sure you write down your prediction *before* performing the experiment.

7. Try the experiment. **Note:** You can bring this paperclip quite close to the floating paperclip, but make sure they don't touch or you may sink the floater.

Question 2-7: Describe your observations. Do they make sense? Explain.

Comments: It should be very clear from the last activity that a paperclip can be turned into a (weak) magnet simply by placing it near the poles of a magnet. This process goes by the descriptive term magnetization. But what is *magnetization*?

The process of magnetization is actually quite difficult to explain on a fundamental level. When an object made from a magnetic material is brought near a magnet, it is turned into a magnet. The closer this object is brought to one of the poles of the magnet, the stronger this effect is. That is, the effect of magnetization is stronger when the object is placed in a stronger magnetic field.

But how and why does this happen? The answer to this question has to do with the quantum-mechanical property of an electron called *spin*.[2] One of the consequences of electron spin is that the electron behaves like a tiny magnet. In many materials, electrons in an atom are paired up with opposite spin so there are no macroscopic magnetic effects. However, in ferromagnetic materials such as iron, unpaired electrons like to point in the same direction. If they are placed in a strong magnetic field, these tiny paired electrons can act together and become a strong magnet.

Of course, like the paperclip, magnets are not truly permanent. After some time, the magnetization will wear off. In fact, one way to demagnetize a "permanent magnet" rapidly is to drop it. This is why your instructor will probably tell you not to drop magnets!

---

[2] Although it is tempting to think of an electron as spinning around, this is not an accurate picture for what spin is. Unfortunately, we cannot give a good picture for what spin is because there is no classical analogy for spin. It is a purely quantum-mechanical effect.

INVESTIGATION 3: CAN MAGNETIC FORCES ACT ON CHARGES?

Maybe magnetic and electrical forces are the same thing, or perhaps they are related in some way. We now know that a magnet exerts a force on another magnet, but can it exert a force on electric charges? In this investigation you will explore this question.

Prediction 3-1: In Lab 1, you saw that static electric charges can be placed on pieces of Scotch Magic© tape by first sticking the tape to a tabletop or on top of another piece of tape, and then pulling it off. What effect do you think a magnet will have on a charged piece of Scotch tape with a net positive or net negative charge? Describe what you think will happen if a magnet is brought near a charged piece of tape.

To test your prediction, you will need

- 2 10-cm-long Scotch Magic© tape pieces
- rod-shaped magnet (with labeled poles)
- non-magnetic (aluminum) conducting rod the same size as the magnet

Activity 3-1: The Magnetic Force Exerted on Static Charges— Charged Scotch Magic© Tape

1. Test to see if there is any force on the electrically charged tape from either pole of your magnet. Do the same test with a non-magnetic conducting rod.

Question 3-1: Summarize your findings.

Question 3-2: If the charged tape is attracted to both the magnetic *and* the non-magnetic rod in the same way, can you conclude that there is any special interaction or force between either of the magnetic poles and the tape?

Question 3-3: Are magnetic attractions the same as electrostatic attractions? Cite evidence for your answer.

Magnetic Forces on Moving Charges

Let's try something unusual. Let's see if a magnet can exert forces on electrical charges that are *moving*.

Prediction 3-2: Does a magnetic field ever exert a force on a charged particle? Will there be a magnetic force on electrons moving in the vicinity of a magnet (in the presence of a magnetic field)?

For this observation you will orient a permanent magnet in various directions relative to a beam of electrons that travels along the axis of an evacuated oscilloscope tube. There is a phosphorescent screen on the front of the tube that shows you where the electrons hit. With your magnet poles (labeled N and S), you can perform a qualitative investigation of the nature of the force, if any, that a magnetic field can exert on the beam of electrons. To test your prediction, you will need:

- demonstration oscilloscope (with open side)
- rod-shaped magnet (with labeled poles)

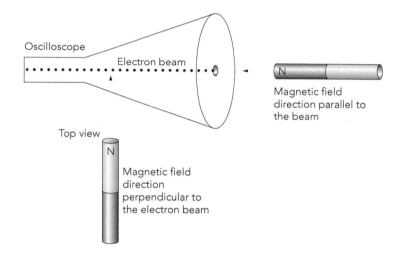

Activity 3-2: The Magnetic Force Exerted on Moving Charges—An Electron Beam in an Oscilloscope

Question 3-4: Is the charge on an electron positive or negative? What is the direction of motion of the electrons in the beam in the diagram above—Right to Left or Left to Right?

1.   Move the north pole of your magnet parallel and then perpendicular to the electron beam in the oscilloscope.

Question 3-5: What is the direction of the displacement (and hence the force on the beam) in each case? Sketch vectors below showing the direction of the magnetic field, the direction of motion of the original electron beam before it was deflected, and the direction of the resultant force on the beam.

Question 3-6: Carefully describe in words the relationship between the direction of motion of the electron beam, the direction of the magnetic field and direction of the force on the electrons. Are any of these perpendicular to each other?

2. With the magnet in a fixed position, observe what happens when the velocity of the electrons is increased. **Note:** You can increase the oscilloscope's beam velocity by turning up the intensity control. *Since the intensity control on a typical oscilloscope only increases the velocity of the electrons by a relatively small amount, look for a subtle change.*

Question 3-7: Describe what happens when the speed of the electrons is increased.

3. With the speed of the electrons constant, observe what happens when the magnetic field at the position of the electrons is increased by moving the magnet closer.

Question 3-8: Describe what happens to the displacement of the electrons when the magnetic field strength is increased.

Question 3-9: Are your observations consistent with the magnitude of the magnetic force being described by $F = qvB$ where q is the charge, v is the speed of the moving charge, and B is the magnitude of the magnetic field? Explain based on your observations.

4. As you should have found, the force magnitude depends on the orientation of the magnetic field $\vec{B}$ relative to $\vec{v}$. Assuming that this dependence is a simple sine or cosine dependence, determine which function it is by making further observations.

Question 3-10: Is it a sine or cosine dependence? Explain your reasoning based on your observations.

Definition of the Vector Cross Product

There is a weird mathematical entity called the *vector cross product*. You may have seen it before. Using very general notation, the cross product of two vectors, $\vec{a}$ and $\vec{b}$ is defined as follows:

(1) *Magnitude:* The magnitude of the cross product is given by ab $\sin\theta$, where θ is the angle between the two vectors.

(2) *Direction:* The cross product, $\vec{c}$ is a vector that lies in a direction perpendicular to both $\vec{a}$ and $\vec{b}$. Its actual direction is given by a right hand rule as follows: Point your fingers in the direction of $\vec{a}$ and rotate them into the direction of $\vec{b}$ through the smaller angle θ. Your thumb then points in the direction of $\vec{c}$ as shown in the figure below.

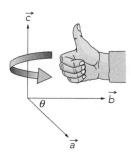

(3) *Notation:* The cross product of two vectors as defined in (1) and (2) is written as follows:

$$\vec{c} = \vec{a} \times \vec{b}$$

Question 3-11: Are your observations in Activity 3-2 consistent with the representation of the magnetic force on a moving charged particle as

$$\vec{F} = q\vec{v} \times \vec{B}$$

Give specific evidence from your observations of the magnitude *and direction* of the force. **Note:** The sign of the charge on the electron needs to be considered.

Comment: The relationship $\vec{F} = q\vec{v} \times \vec{B}$ is known as the Lorentz force equation.

INVESTIGATION 4: MAGNETIC FORCE ON CURRENT CARRYING WIRES

You demonstrated in Investigation 3 that permanent magnets do not exert forces on the static charges on a piece of charged Scotch tape. On the other hand, magnetic fields from a permanent magnet can bend a beam of electrons.

You can now use the Lorentz force equation to predict *qualitatively* the nature of the force on a non-magnetic wire when it is placed between the poles of a strong permanent magnet so that part of the wire is in a fairly strong magnetic field.

Activity E4-1: Forces on a Wire

Prediction E4-1: Predict the force when the wire carries no current.

Prediction E4-2: Predict the force when the wire carries current, and what will happen if the current is reversed.

To test your predictions, and examine the effects of a magnetic field on a wire you will need:

- 6 V lantern battery or power supply
- 2 alligator clip leads, 20 cm
- 2 alligator clip leads, 10 cm
- SPST switch
- strong horseshoe magnet

1. Set up the circuit shown in the diagram below, *but do not close the switch.*

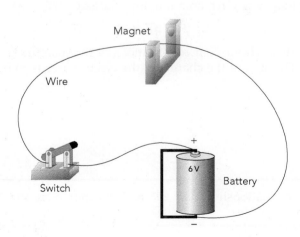

2. Make observations to test your predictions. **Warnings:** (1) Keep the alligator clip leads away from the magnet's poles as they have iron in them. (2) Do not leave the switch on for long or the battery will go dead quickly.

Question E4-1: What happens when there is no current flowing in the wire (switch open)? Describe your actual observations. If there is a force on the wire, note the *direction* of the force relative to the direction of the magnetic field.

Question E4-2: What happens when there is a current flowing in the wire (switch closed <u>briefly</u>)? Describe your actual observations. If there is a force on the wire, note the *direction* of the force relative to the directions of the magnetic field and the current. What happened when the current was reversed?

Question E4-3: Are your observations consistent with the Lorentz force equation. Be specific in basing your answer on your observations, and don't forget direction.

Name_____   Date_____   Partners_____

HOMEWORK FOR LAB 9
MAGNETISM

1. Can you find isolated north or south magnetic poles, or do they always come in pairs? Is this the same or different for + and − electric charges?

2. Describe two similarities and two differences between electric and magnetic field lines. (Consider such things as where they originate and terminate, how they are related to the direction and strength of the field, whether they are closed curves or lines, and whether there's anything you can say about their flux through a closed surface.)

3. In each of the diagrams below, describe in which direction the charged particle will be deflected under the influence of the magnetic field of the magnet shown.

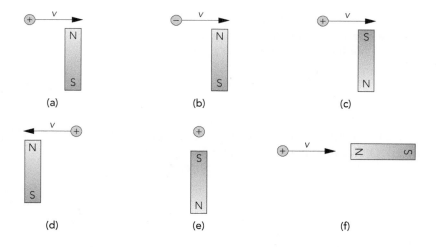

4. In each case in (3), explain how your answer agrees with the Lorentz Force equation, $\vec{F} = q\vec{v} \times \vec{B}$

    a.

    b.

    c.

    d.

    e.

    f.

5. A charged particle with charge $+6.0 \times 10^{-9}$ C moves at a speed of 2.5×10^6 m/s in a magnetic field 0.50 T as shown below.

   Find the magnitude of the force on this particle, and specify its direction.

6. A wire carries current into the paper, between the poles of a horseshoe magnet, as shown on the right in (a). Indicate the direction of the force on the wire.

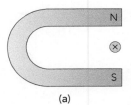

   (a)

   A wire carries current out of the paper, between the poles of a horseshoe magnet, as shown on the right in (b). Indicate the direction of the force on the wire.

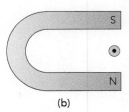

   (b)

7. Explain your answers in (6) using the Lorentz Force equation.

    a.

    b.

8. Explain using the Lorentz Force equation why a charged particle moving at a constant speed in a direction perpendicular to a uniform magnetic field moves in a circle.

9. Sketch the subsequent path of the electron that is instantaneously moving in the direction shown in the diagram below.

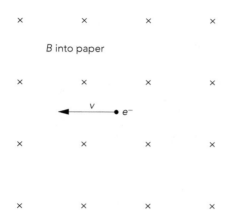

Pre-Lab Preparation Sheet for
Lab 10—Electromagnetism

(Due at the beginning of lab)

Directions:
Read over Lab 10 and then answer the following questions about the procedures.

1. What will you use to determine the magnitude and direction of the magnetic field produced by a current-carrying wire in Investigation 1?

2. What is the purpose of the meter in Activity 2-1?

3. What do you predict will happen in Activity 2-1 when the magnet is sitting with its North pole in the middle of the coil?

4. What do you predict will happen in Activity 2-2 when the coil is pulled away from the North pole of the magnet?

5. What do the movies in Investigation 3 show? What do the graphs show?

LAB 10:
ELECTROMAGNETISM

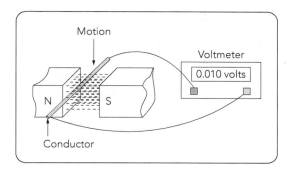

When the magnet is introduced, the needle at the galvanometer is deflected in a certain direction; but being in, whether it be pushed quite through or withdrawn, the needle is deflected in a direction the reverse of that previously produced. When the magnet is passed in and through at one continuous motion, the needle moves one way, is then suddenly stopped, and finally moves the other way.

Michael Faraday (1791–1867)

OBJECTIVES

- To observe the magnetic field produced by electric current flowing in a straight wire and invent a right hand rule to determine the direction of the associated field lines.

- To discover under what circumstances a magnetic field can induce a voltage in a coil of wire.

- To relate the voltage and current induced in a coil of wire to the rate of change of magnetic flux passing though it.

- To verify Faraday's law of electromagnetic induction.

- To verify Lenz's law of electromagnetic induction.

OVERVIEW

In the last lab you observed that permanent magnets can exert forces both on freely moving charges and on electric currents in conductors. We have postulated the existence of a mathematical entity called the magnetic field in order to introduce the Lorentz force law as a way of describing mathematically the nature of the force that a magnet can exert on moving electrical charges. Newton's Third law states that whenever one object exerts a force on another object, the second object exerts an equal and opposite force back on the first object. Thus, if a magnet exerts a force on a current-carrying wire, shouldn't the wire exert an equal and opposite force back on the magnet? It is not unreasonable to speculate that currents and moving charges exert these forces by *producing magnetic fields themselves*. One of the agendas for this lab is to investigate the possibility that an electric current can produce a magnetic field.

This line of argument based on Newton's Third law and its symmetry can lead us into even deeper speculation. If charges have electric fields associated with them, then moving charges can be represented mathematically by changing electric fields. Thus, we can say that changing electric fields are the cause of magnetic fields. This leads inevitably to the following question: If this is so, then, by symmetry, *can changing magnetic fields cause electric fields?*

You will explore three phenomena in this lab: (1) The nature of the magnetic field produced by a long straight current-carrying wire; (2) Faraday's law of electromagnetic induction that describes how changing magnetic fields produce electric fields, emfs and currents; and (3) Lenz's law that describes the sense or direction of the emfs and currents produced by electromagnetic induction.

Faraday's law lies at the heart of the study of electricity and magnetism, and electromagnetic induction is the basis for the production of the electric power on which we depend.

INVESTIGATION 1: MAGNETISM FROM ELECTRICITY

The Magnetic Field Near a Current-Carrying Wire

In 1819 during a lecture demonstration a Danish physicist, H.C. Oersted, placed a current carrying wire near a compass needle. Although Oersted predicted that the current would cause a force to be exerted on the compass needle, the details of the result surprised him. What do you predict will happen?

Prediction 1-1: Do you expect to detect a magnetic field in the vicinity of a straight current-carrying wire? Why?

Prediction 1-2: If your Prediction 1-1 is yes, what do you think will be the direction of the magnetic field near the wire? What do you think will happen to the direction of the magnetic field if the direction of the current in the wire is reversed?

Prediction 1-3: Do you expect the magnitude of the field to increase, decrease, or stay the same as the distance from the wire increases. Why?

Prediction 1-4: What to you think will happen to the magnitude of the magnetic field if the current is reduced?

To test your predictions, you can repeat some of Oersted's observations and study the pattern of magnetic field lines in a plane perpendicular to a long straight conductor that is carrying current. To do this you can use:

- 2   6 V batteries
- 3   30 cm lengths of wire with alligator clip leads
- switch
- 5 ohm resistor
- non-magnetic (aluminum) ring stand with clamps and rods
- 20 cm × 20 cm piece of cardboard (or acrylic plate) with a hole
- small compass
- magnetic field sensor

Activity 1-1: Magnetic Field of a Straight Current-Carrying Wire

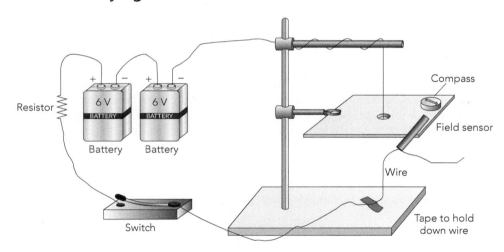

Figure 10-1: Apparatus for repeating Oersted's observations on the magnetic field produced by a current showing correct orientations of field sensor and compass.

1. Wire the two 6V batteries, the resistor, the switch, and wires in series as shown in Fig. 10-1. The center wire can be poked through a hole in the piece of cardboard (or acrylic) that's mounted perpendicular to the wire.

2. First wire the batteries as shown so that positive current flows through the wire from bottom to top.

3. Use the compass to determine the direction of the field at various points around the wire. *To save the batteries, only connect them when you're making observations.* **Note:** Before the current is switched on, make sure your compass is pointing north and freely swinging.

4. Use the diagram below (top view) to map out the magnetic field directions with arrows. **Note:** This is a top view, looking down the wire. Try placing the compass at different points around imaginary circles centered on the wire.

Question 1-1: What is the shape of the magnetic field lines? Be precise. How do your results agree with your Predictions 1-1 and 1-2?

5. Reverse the current, and repeat your measurements.

Question 1-2: What happens to the direction of the magnetic field when the current flows down through the wire instead of up? Be precise. How do your results agree with your Prediction 1-2?

6. Wrap the fingers of your right hand around the wire with your thumb pointing in the direction of the current, and figure out a rule for using your right hand to determine the direction of the magnetic field surrounding the wire. **Hint:** (1) Review the definition of magnetic field direction from Lab 9, and (2) pay attention to the direction your fingers are pointing!

Question 1-3: Describe your *right hand rule* for determining the direction of the magnetic field, and explain how it describes the directions of the field correctly for the current flowing up and down through the wire. Be precise.

7. Open the experiment configuration file **Magnetic Field (L10A1-1)** to set up the magnetic field sensor, and, if necessary, **zero** it *with it far away from any magnets.*

8. Probe the field strength around the wire. **Note:** The sensor should always be held so that the field lines are perpendicular to its face. Rotate the sensor a bit to find the maximum field at each position.

9. First hold the field sensor in a horizontal position far away from the wire to see what zero field looks like. Then move the sensor close to the wire, and slowly move it away from the wire while observing whether the magnetic field is increasing, decreasing or remaining the same. Hold the switch in the circuit down just long enough to make your observations.

Question 1-4: How does the field strength vary as you move away from the wire? Increase? Decrease? Or stay the same? Indicate your results on the diagram to the right using the symbols SS=strongest, S=strong, W=weaker and WW= weakest. How does this agree with your Prediction 1-3?

11. Slowly move the sensor to measure the field at various distances on the opposite side of the wire.

Question 1-5: Does the field vary in the same way on both sides of the wire? Explain.

12. Measure the field strength around the wire again. *Remember that the sensor should always be held so that field lines are perpendicular to its face. With the sensor at one position where the field strength* is relatively large, remove one of the batteries from the circuit to reduce the current through the wire.

Question 1-6: What happens to the strength of the magnetic field at a given distance from the wire when the current in the wire is decreased? How does your result agree with your Prediction 1-4?

Question 1-7: Summarize your observations of the magnetic field produced by a long current-carrying wire below:

Dependence of magnitude on distance from the wire:

Direction of field lines: (How can the right hand rule be used to determine this?)

Dependence of the field magnitude on the current flowing through the wire:

INVESTIGATION 2: INTERACTIONS BETWEEN A MAGNET AND A COIL

You can investigate the effects of a magnetic field on the current in a coil of wire with the following:

- strong rod-shaped magnet

- coil of wire with large number of turns

- an analog galvanometer, voltmeter or ammeter

Under what circumstances can a voltage (emf) be induced in a wire by a magnetic field? Before you make any observations with a magnet and coil, consider the following prediction.

Prediction 2-1: Check which experiments you think will produce a voltage across the coil (or current through it).

_____1. A magnet is located in the middle of the coil and then a meter is connected to the coil.

_____2. While a meter is connected to the coil, the magnet is pulled out of the coil.

_____3. While the coil is connected to the meter, the magnet is pushed into the coil.

Test your predictions.

Activity 2-1: Magnet and Coil

1. Place the magnet in the middle of the coil.

2. Connect the coil to the meter, and make observations to test each of the three predictions.

Question 2-1: In which, if any, of your observations, did you observe an electrical effect in the coil? Describe what you observed.

Question 2-2: If you observed an electrical effect in more than one of your experiments, compare your observations. What was the same and what was different?

3. Repeat all of your observations with the opposite pole of the magnet facing the coil.

Question 2-3: With the pole of the magnet switched, describe what was the same as in Question 2-1 and what was different for each observation you made.

Are there any other ways to produce electrical effects in the coil? This time, suppose that the magnet is at rest and the coil is moved. You will need

• something to hold the magnet in place, such as a lab stand.

Prediction 2-2: Place a check next to the observations that you think will produce electrical effects in the coil of wire.

_____1. The coil starts with one of the magnet's poles at its center and is pulled away from the magnet.

_____2. The coil starts with the magnet a small distance away from its center, and is moved toward the magnet.

_____3. Neither of the above.

Prediction 2-3: What do you predict will happen if you repeat the above observations with the opposite pole of the magnet facing the coil?

Test your predictions.

Activity 2-2: Coil Moving Near a Magnet

1. Mount the magnet close to the coil so that the magnet can't move.

2. With the coil connected to the meter in exactly the same way as in Activity 2-1, make observations to test Prediction 2-2. Start or end with one pole of the magnet near the center of the coil.

Question 2-4: In which, if any, of your observations, did you observe an electrical effect in the coil? Describe what you observed. Did your observations agree with Prediction 2-2?

Question 2-5: If you observed an electrical effect in more than one of your experiments, compare your observations. What was the same and what was different?

3. Reverse the poles of the magnet, and repeat your observations.

Question 2-6: What was the effect of reversing the poles of the magnet?

Question 2-7: Now compare your observations to those in Activity 2-1. Did any of the experiments result in the same observations? If so, which ones?

Let's explore what is the same and what is different about the observations in Activities 2-1 and 2-2. In the former activity, the magnet is moved. In the latter the coil is moved.

Question 2-8: If you observed that moving the coil in some way and moving the magnet in some way resulted in the same electrical effect, describe these two experiments, and speculate what is the same in both experiments.

Activity 2-3: The Magnetic Field of a Rod-shaped Magnet

1. Based on your observations in Lab 9, sketch some magnetic field lines for a rod-shaped magnet. Indicate in your diagram places where the magnetic field is strong and places where it is weaker. Use the following symbols: SS= strongest, S=strong, W=weaker, WW=weakest.

Question 2-9: When you move the magnet away from the coil, what happens to the magnetic field in the vicinity of the coil? When you move the coil away from the magnet, what happens to the magnetic field in the vicinity of the coil? Explain based on your observations.

Question 2-10: What do you think is responsible for inducing electrical effects in a coil of wire, the presence of a magnetic field or the presence of a changing magnetic field? Explain based on your sketch above.

Question 2-11: When the magnet is left sitting at rest in the center of the coil, is the magnetic field in the vicinity of the coil changing? How is your observation of this in Activity 2-1 consistent with your answer to Question 2-10?

Magnetic Flux

You have observed that when the magnetic field in the vicinity of a coil of wire changes, an emf (voltage) is induced across the coil. It turns out that the induced emf is not directly related to the change in the field, but rather to the change in *magnetic flux* through the coil. Magnetic flux is defined in the same way as electric flux. From Lab 2, and the definition of electric flux, you know that the magnetic flux through a surface is defined as

$$\Phi_M = \Sigma \Delta \Phi_M = \Sigma \, B \cos \theta \, \Delta A \text{ [flux through a surface]}$$

Here, ΔA is the magnitude of the area of a piece of the surface where the field has magnitude B, and θ is the angle between the normal to the surface and the direction of the field.

So the magnetic flux through a surface is the sum over the whole surface of the products of the magnetic field component perpendicular to the surface ($B \cos \theta$) at each location and a small area element at each point (ΔA).

Question 2-12: Based on this definition, describe three ways that you can produce a changing magnetic flux through the surface of a coil.

(1)

(2)

(3)

Question 2-13: For each of your observations in Activities 2-1 and 2-2 where you saw an emf induced in the coil, describe which of the methods in Question 2-12 was used to produce the changing flux.

Faraday's Law

How exactly does the emf induced in a coil of wire depend on the change in the magnetic flux through the coil? The answer is found theoretically in Faraday's law. Before writing down this law, you will examine this question semi-quantitatively. Then, in Investigation 3 you will do a more quantitative analysis.

First several predictions.

Prediction 2-4: Suppose you start with the North pole of the magnet in the middle of the coil, and the coil connected to the meter. If you pull the magnet straight out from the coil, will the maximum induced emf be larger when you pull it out slowly or when you pull it out rapidly? Or will you observe the same maximum emf in both cases?

Prediction 2-5: Suppose you hold the magnet with its North pole close to the opening in the coil. You then rotate the coil around an axis perpendicular to the table. Will there be an induced emf? If yes, will the maximum value of the emf during each rotation depend on the angular speed of the coil? How?

To test these predictions, you will need the same materials that you used for Activities 2-1 and 2-2.

Activity 2-4: Dependence of Induced EMF on Speed of Moving Magnet

1. Set up the coil, meter and magnet as in Activity 2-1. Start with the North pole of the magnet at the center of the coil, and then pull the magnet out quickly.

2. Repeat the observation while pulling the magnet out slowly.

Question 2-14: Was there any difference in your observations pulling out the magnet quickly and slowly? Describe carefully.

Question 2-15: Compare the change in magnetic flux through the coil when the magnet is pulled out quickly and slowly, but otherwise in the same manner. Is the change in flux the same in both cases or different? (This is a thought question; you do not actually observe the change in flux. Think about whether the flux was the same at the beginning for both cases, and the same at the end for both cases.)

Question 2-16: In Question 2-14, if you reported a difference in induced emf when the speed of the coil was different, what do you conclude? Does the induced emf depend on the *change* in the magnetic flux through the coil or the *rate of change* of the flux? Explain, based on your observations.

Extension 2-5: How Does Induced EMF Depend on the Speed of a Coil's Rotation?

1. Position the coil so that the North pole of the magnet is close to and perpendicular to the center of the face of the coil. The meter should be connected to the coil, as in Activity 2-1.

2. Rotate the coil quickly around an axis perpendicular to the table, and observe any reading on the meter.

3. Repeat (2), rotating the coil in the same manner but slowly.

Question E2-17:   Was there any induced emf in (2)? If so, explain why the flux through the coil was changing.

Question E2-18:   Was there any difference in your observations rotating the coil quickly and slowly? Describe carefully.

Question E2-19:   Compare the change in magnetic flux through the coil when the coil is rotated quickly and slowly, but otherwise in the same manner. Is the change in flux the same in both cases or different?

Question E2-20:   If your answer to Question E2-18 was that the induced emf was different, what do you conclude? Does the induced emf depend on the *change* in the magnetic flux through the coil or the *rate of change* in the magnetic flux. Explain, based on your observations.

Comment: According to Faraday's law, at any moment the magnitude of the induced emf (denoted ε) across a conducting coil with N turns is equal to N times the negative rate of change of the magnetic flux Φ_M passing through the area enclosed by the coil as described in the following equation:

$$\varepsilon = -N\frac{\Delta \Phi_M}{\Delta t}$$

Question 2-21:   Describe how Faraday's law is consistent with your observations in Activity 2-4 and Extension 2-5. For now, don't worry about the meaning of the minus sign in the equation.

INVESTIGATION 3: FARADAY'S LAW AND ELECTROMAGNETIC INDUCTION—A MORE QUANTITATIVE LOOK

In this investigation you will be asked to use Faraday's law to explain why the current induced in a coil and thus the voltage across it changes the way it does when the North or South pole of a rod shaped magnet is moved into and then back out of the center of the coil. No calculations are required in this investigation, just thinking!

Figure 10-2 shows sequential photos of how the magnet will be moved. Rather than moving the magnet yourself, you will have four movies, A, B, C and D, that show someone moving the magnet in different ways through the coil.

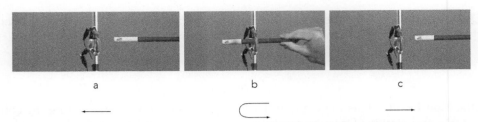

a   b   c

Figure 10-2: The north pole of a rod magnet is: (a) moving into the center of a coil of wire; (b) reversing direction; and then (c) moving out of the coil. The coil's voltage vs. time will be recorded using a computer interfaced voltage sensor.

Variation of Magnetic Field Passing though a Coil in the Vicinity of a Rod-Shaped Magnet: Figure 10-3 shows the direction of the magnetic field lines associated with a rod-shaped magnet along with the area vector associated with the coil. By convention, if the coil's area vector and the field line are in the same direction, the flux passing through the coil is positive. If the field line direction and the area vector are opposite in direction, then the flux passing through the coil is negative.

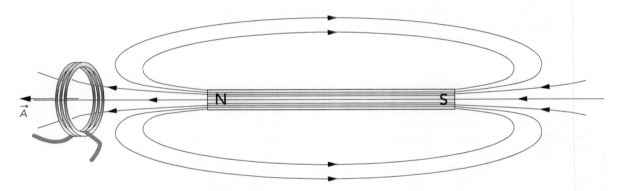

Figure 10-3: External field lines point away from the North pole and toward the South pole.

Question 3-1: How do the lines in Figure 10-3 compare to those you drew in Activity 2-3? Give a detailed comparison.

Recall that change in flux is the difference between the final value of the flux and its initial value.

Prediction 3-1: If the North pole of the magnet shown in Fig. 10-3 is moved toward the left from its initial position until it is just at the center of the coil, will the sign of the change in flux, $\Delta\Phi = \Phi_{M,\,final} - \Phi_{M,\,initial}$, through the coil be negative, positive or zero? Explain your prediction.

Prediction 3-2: What happens to the *absolute value* of the *rate of change* of the flux, $|\Delta\Phi_M/\Delta t|$ through the coil, if the magnet is moved more rapidly through the coil? Explain your prediction.

You have just started a new engineering job to help design a miniature generator. Your supervisor asks you to start by examining data she took as a rod-shaped magnet was moved back and forth in a direction perpendicular to the coil, as shown in Figure 10-2.

Activity 3-1: Practice Observing Data for Induced EMF

You will begin by opening the experiment file **Sample Movie (L10A3-1).** This file allows you to replay a movie of the motion of the magnet, and also includes a graph of the measured emf. (Your boss had already taken video data of the rod's motion to create a graph of the x-component of the rod's leading edge and its velocity vs. time.)

1. **Press the start button** on the replay window to observe the movie of how the magnet was moved.

2. The emf, ε, induced in the coil (monitored by a voltage sensor interfaced with a computer) is also displayed. You can replay this by selecting **Latest** in the drop-down menu.

3. You can see how ε changes over time. Think about how the relative velocities, polarity, and magnetic flux passing through the coil should affect the measured emf.

Question 3-2: Look at the motion of the magnet, and the graph of the induced emf, and compare these to your Prediction 3-2. Does the observed induced emf agree with your prediction? Explain.

Activity 3-2: Analyzing Different Situations

Your supervisor gives you the four emf vs. time graphs, Graphs 1-4, shown below. Each was created with the motion of the magnet relative to the coil shown in one of the four movies, A, B, C or D. **Note:** In all four movies the rod-shaped magnet starts by moving to the *left* before it encounters the coil.

1. To view the movies A, B, C or D, open the experiment files **Movie A (L10A 3-2A), Movie B (L10A3-2B), Movie C (L10A3-2C),** or **Movie D (L10A3-2D),** respectively.

2. Examine these movies and use your understanding of Faraday's law to deduce which movie corresponds to which data set.

3. In the space below, draw a line from each data set to its corresponding movie.

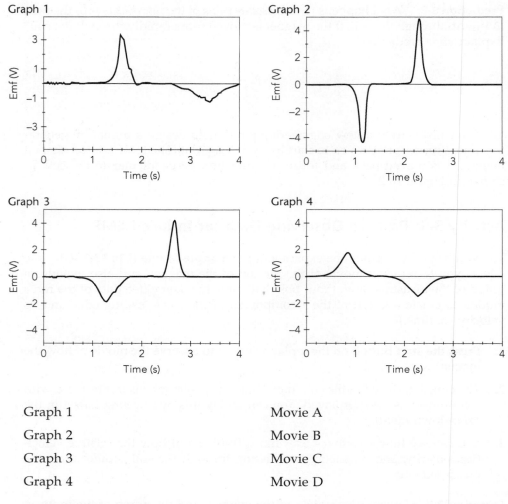

Graph 1

Graph 2

Graph 3

Graph 4

Movie A

Movie B

Movie C

Movie D

Question 3-3: Explain in detail why you associated Graph 1 with the movie you chose.

Question 3-4: Explain in detail why you associated Graph 2 with the movie you chose.

Question 3-5: Explain in detail why you associated Graph 3 with the movie you chose.

Question 3-6: Explain in detail why you associated Graph 4 with the movie you chose.

INVESTIGATION 4: LENZ'S LAW AND ELECTROMAGNETIC INDUCTION

In Investigation 3, you have seen in the movies and graphs that the induced emf can be either positive or negative, and, therefore, the induced current can flow in different directions depending on how the magnet is moved relative to the coil. We did not attempt to use Faraday's law to explain the direction of the emf. Here we will explore this further.

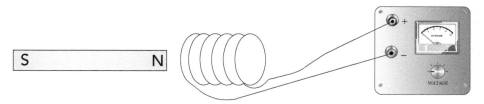

Figure 10-4: Magnet, coil of wire and meter.

Prediction 4-1: If there is a current induced in the coil by movement of the magnet, and the meter reads a positive current, describe how the current is flowing around the loops of the coil: clockwise or counterclockwise as viewed from right.

Prediction 4-2: Consider the right hand rule for determining the direction of the magnetic field produced by a current flowing in a wire. (Refer back to Questions 1-3 and 1-7.) If a current is flowing counterclockwise in the loops of the coil (as viewed from the right), will the magnetic field produced by the coil be toward the right or toward the left?

Prediction 4-3: What is the general direction of the magnetic field on the right (N) side of the rod-shaped magnet, caused by the magnet: toward the right or toward the left?

Prediction 4-4: If the rod-shaped magnet is moved toward the coil, does the magnetic flux through the coil increase or decrease?

You can now use these predictions to discover something very interesting about the direction of flow of the induced current. You will need

- a coil of wire with ends exposed so that the direction of the winding can be determined
- a strong rod-shaped magnet
- an analog galvanometer, ammeter or voltmeter

Activity 4-1: Lenz's Law

1. Connect the coil to the meter as shown in Figure 10-4. Be sure that the direction of windings on the coil, and connections to the meter, are exactly as in Figure 10-4.

2. Move the N pole of the magnet quickly toward the center of the coil, and note whether the current is positive or negative.

Question 4-1: Is the current positive or negative? Does this represent a clockwise or counterclockwise current (viewed from the right) in the loops of the coil?

Question 4-2: Based on your analysis for Prediction 4-2, the induced current in the coil produces a magnetic field in which direction? Explain.

Question 4-3: When you move the N pole of the magnet toward the center of the coil, do you increase or decrease the flux through the coil toward the right? Explain.

Question 4-4: Is the flux from the magnetic field produced by the *induced* current (Question 4-2) toward the right or left? Explain. Does this flux add to or subtract from the change in flux from the movement of the magnet (Question 4-3)? Explain.

> **Comment:** Lenz's law says that when electromagnetic induction takes place, the induced current (or voltage) has a direction (or sense) that tends to oppose the effect that produced it.

Question 4-5: What causes the induced current in this experiment? **Hint:** What is causing the flux to change? Does the induced current flow in a direction so that it opposes this change in flux?

Prediction E4-5: Suppose you reverse the magnet so that the South pole is on the right. Also, you begin with the South pole close to the center of the coil, and then move it quickly away from the coil.

1. Is the flux through the coil caused by the magnet toward the right or left?

2. What happens to this flux when the magnet is moved away?

3. According to Lenz's law, in what direction should the magnetic field produced by the induced current be?

4. In which direction around the loops of the coil—clockwise or counterclockwise—must the induced current flow to produce the field predicted in (3)?

5. Will the meter reading be positive or negative?

Extension 4-2: Another Test of Lenz's Law

1. Reverse the poles of the magnet.
2. Use the same coil connection as in the previous activity.
3. Start with the S pole near the center of the coil and pull the magnet away quickly.

Question E4-5: Is the current positive or negative? Did this agree with your prediction? Explain.

Comment: Lenz's law is a powerful method for determining the direction or sense of electromagnetically induced effects, no matter how they are produced.

HOMEWORK FOR LAB 10
ELECTROMAGNETISM

1. In the diagram on the right, a wire is carrying a current. The magnetic field above the wire is into the paper, and the magnetic field below the wire is out of the paper, as shown. (An x indicates a field line into the paper and a • indicates a field line out of the paper.) What is the direction of current flow in the wire? Explain your answer based on your observations in this lab and the right hand rule.

2. Is the magnetic field magnitude larger at point 1 or point 2? Explain based on your observations in this lab.

3. You have a rod-shaped magnet and also a coil of wire connected to a voltmeter. Under which of the following circumstances will there be a voltage induced across the coil? If you predict that a voltage is induced, explain why. If you think there will be no voltage, explain why not.

   a. The magnet is sitting near the coil, and the coil is rotated so that its face is no longer perpendicular to the magnet.

   b. The magnet is sitting near the coil, and the coil is moved quickly away from the magnet.

   c. The magnet is sitting at rest with its North pole in the middle of the coil.

   d. The magnet is sitting with its North pole in the middle of the coil and is pulled away from the coil.

4. If there is a changing flux through the coil in any of the circumstances described in (3), explain what is causing the flux to change.

    a.

    b.

    c.

    d.

5. Calculate the magnetic flux through the coil in each case. The magnetic field is 2.0 Tesla, and the area of the face of the coil is 0.25 m².

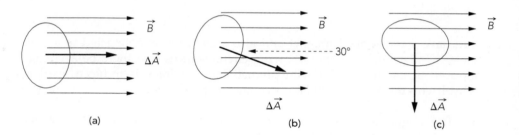

(a)   (b)   (c)

6. The North pole of a rod-shaped magnet is near the center of a coil of wire connected to an ammeter as shown below. The coil is pulled quickly away from the magnet. Use Lenz's law to predict whether the reading on the meter will be positive or negative. Explain your answer carefully.

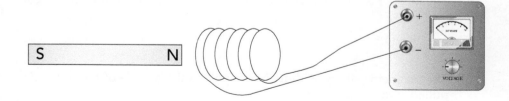

7. A rod-shaped magnet is held with one end a small distance away from the center of a coil of wire attached to a voltmeter, as shown to the right. The magnet is aligned with the area vector of the coil. Shown below are two possible graphs of the voltmeter reading as a function of time. In each case, describe how the magnet could be moved to produce the emf described by the graph. Specify which pole must be facing the coil, and discuss the direction and speed of movement.

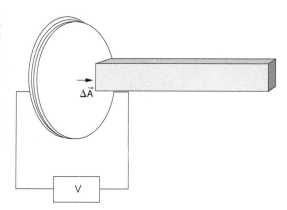

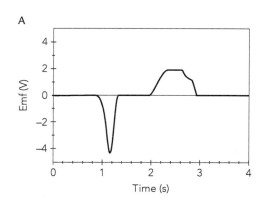

A

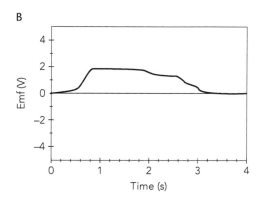

B

APPENDIX: REALTIME PHYSICS ELECTRIC CIRCUITS EXPERIMENT CONFIGURATION FILES

Listed below are the settings in the *Experiment Configuration Files* used in these labs. These files are available from Vernier Software and Technology for *Logger Pro* software (Windows and Macintosh) and from PASCO for *Data Studio* (Windows and Macintosh). They are listed here so that the user can set up files for any compatible hardware and software package.

Experiment File	Description	Data Collection	Data Handling	Display
Coulomb's Law (L01A2-3)	Displays movie with interactive video analysis.	Manual using video analysis.	NA	Movie with video analysis tools.
Force vs. r (L01A2-4)	Displays data table and axes for Force vs. r. Data can be entered into the table, and an active graph results.	Data are entered manually by double clicking on spaces in the data table.	NA	One set of axes with appropriate scales in N and m.
Flux vs. Angle (L02A1-3)	Displays data table and axes for Flux vs. Angle. Data can be entered into the table, and an active graph results.	Data are entered manually by double clicking on spaces in the data table.	NA	One set of axes with appropriate scales in W and Degrees.
Measuring Force (L03A1-3)	Displays and graphs Force vs. time.	20 points/sec.	Use Analysis Feature to find mean force.	One set of axes for Force in N vs. time in s.
Current Model (L04A1-5)	Displays and graphs Current 1 and Current 2 vs. time.	25 points/sec Current sensors 1 and 2 Digital and graphical display of inputs	NA	Two sets of graph axes with lines Current: -0.6 to $+0.6$ A Time: 0–10 sec
Two Voltages (L04A2-4a)	Displays and graphs Voltage 1 and Voltage 2 vs. time.	25 points/sec Voltage sensors 1 and 2 Digital and graphical display of inputs	NA	Two sets of graph axes with lines. Voltage: -5 to $+5$ V Time: 0–10 sec
Current and Voltage (L04A2-4b)	Displays and graphs Voltage 1 and Current 2 vs. time.	25 points/sec Voltage sensor 1 and Current sensor 2 Digital and graphical display of inputs	NA	Two sets of graph axes with lines. Voltage: -5 to $+5$ V Current: -0.6 to $+0.6$ A Time: 0–10 sec
Two Currents (L05A1-2)	Displays and graphs Current 1 and Current 2 vs. time.	25 points/sec Current sensors 1 and 2 Digital and graphical display of inputs	NA	Two sets of graph axes with lines. Current: -0.6 to $+0.6$ A Time: 0–10 sec
Batteries (L06A1-2)	Displays and graphs Voltage 1 and Voltage 2 vs. time.	25 points/sec Voltage sensors 1 and 2 Digital and graphical display of inputs	NA	Two sets of graph axes with lines. Voltage: -5 to $+5$ V Time: 0–10 sec

Experiment File	Description	Data Collection	Data Handling	Display
Internal Resistance (L06A2-2)	Displays and graphs Voltage 1 and Current 2 vs. time.	25 points/sec Voltage sensor 1 and Current sensor 2 Digital and graphical display of inputs	NA	Two sets of graph axes with lines. Voltage: −5 to +5 V Current: −0.6 to +0.6 A Time: 0–10 sec
Ohm's Law (L06A3-1)	Displays and graphs Current 1 and Voltage 2 vs. time.	25 points/sec Current sensor 1 and Voltage sensor 2 Digital and graphical display of inputs	NA	Two sets of graph axes with lines. Voltage: 0 to +3 V Current: 0 to +0.5 A Time: 0–30 sec
Dependence of C (L08A1-2)	Displays data table and axes for Capacitance vs. Area (or Separation). Data can be entered into the table, and an active graph results.	Data are entered manually by double clicking on spaces in the data table	NA	One set of graph axes. Capacitance: 0–1.0 nF Area: 0–0.5 m^2 Seperation: 0–100 mm
Capacitor Decay (L08A3-1)	Displays and graphs Current 1 and Voltage 2 vs. time.	25 points/sec Current sensor 1 and Voltage sensor 2. Digital and graphical display of inputs	NA	Two sets of graph axes with lines. Voltage: −6 to +6 V Current: −0.6 to +0.6 A Time: 0–20 sec
Measuring Mag. Field (L09E1-7)	Digital display of Magnetic Field	NA	NA	Digital display.
Magnetic Field (L10A1-1)	Digital display of Magnetic Field	NA	NA	Digital display.
Sample Movie (L10A3-1)	Displays movie, and graphical display of Emf vs. time.	NA	NA	Movie and graph of Emf vs. time.
Movie A (L10A3-2A)	Displays movie, and graphical display of Emf vs. time.	NA	NA	Movie and graph of Emf vs. time.
Movie B (L10A3-2B)	Displays movie, and graphical display of Emf vs. time.	NA	NA	Movie and graph of Emf vs. time.
Movie C (L10A3-2C)	Displays movie, and graphical display of Emf vs. time.	NA	NA	Movie and graph of Emf vs. time.
Movie D (L10A3-2D)	Displays movie, and graphical display of Emf vs. time.	NA	NA	Movie and graph of Emf vs. time.